Springer Theses

Recognizing Outstanding Ph.D. Research

Aims and Scope

The series "Springer Theses" brings together a selection of the very best Ph.D. theses from around the world and across the physical sciences. Nominated and endorsed by two recognized specialists, each published volume has been selected for its scientific excellence and the high impact of its contents for the pertinent field of research. For greater accessibility to non-specialists, the published versions include an extended introduction, as well as a foreword by the student's supervisor explaining the special relevance of the work for the field. As a whole, the series will provide a valuable resource both for newcomers to the research fields described, and for other scientists seeking detailed background information on special questions. Finally, it provides an accredited documentation of the valuable contributions made by today's younger generation of scientists.

Theses are accepted into the series by invited nomination only and must fulfill all of the following criteria

- They must be written in good English.
- The topic should fall within the confines of Chemistry, Physics, Earth Sciences, Engineering and related interdisciplinary fields such as Materials, Nanoscience, Chemical Engineering, Complex Systems and Biophysics.
- The work reported in the thesis must represent a significant scientific advance.
- If the thesis includes previously published material, permission to reproduce this must be gained from the respective copyright holder.
- They must have been examined and passed during the 12 months prior to nomination.
- Each thesis should include a foreword by the supervisor outlining the significance of its content.
- The theses should have a clearly defined structure including an introduction accessible to scientists not expert in that particular field.

More information about this series at http://www.springer.com/series/8790

Michael Esseling

Photorefractive Optoelectronic Tweezers and Their Applications

Doctoral Thesis accepted by
University of Münster, Germany

Springer

Author
Dr. Michael Esseling
Institute of Applied Physics
University of Münster
Münster
Germany

Supervisor
Prof. Cornelia Denz
Institute of Applied Physics
University of Münster
Münster
Germany

ISSN 2190-5053 ISSN 2190-5061 (electronic)
ISBN 978-3-319-38497-9 ISBN 978-3-319-09318-5 (eBook)
DOI 10.1007/978-3-319-09318-5

Springer Cham Heidelberg New York Dordrecht London

Softcover reprint of the hardcover 1st edition 2015

Printed on acid-free paper

Springer is part of Springer Science+Business Media (www.springer.com)

Supervisor's Foreword

Light is not only a manifestation of nature's beauty, but also a versatile, contact-free tool to influence small micro-and nanoscale particles. Since the discovery that light can exert small forces on microscopic objects more than 40 years ago, the field of optical micro manipulation has provided the mature toolkit of optical trapping or optical tweezers. It has nowadays developed into one of the most innovative cutting-edge technologies, and has produced a multitude of interdisciplinary applications in cell biology or nanotechnology. In chemistry and microfluidics, optical forces may remotely actuate micro-components and perform micro-reactions.

Using light to structure nano- or micromatter has become especially popular since it bridges the gap between techniques of controlled top-down approaches and bottom-up self-assembly. It potentially covers a large range of sizes and can be nicely shaped and modified by a multitude of devices, such as spatial light modulators. While optical tweezers are well-suited to realize assemblies of up to a 100 particles, the field of optically-mediated particle trapping for high-throughput applications has been enriched by the invention of opto-electronic tweezers, where optical patterns are converted into electrical forces on a photoconductive surface. This approach combines the flexibility of optical tweezers with the significantly larger forces provided by dielectrophoresis. However, in order to exploit its potential to its full extent, the conversion from optical to electrical forces needs to be mediated by nonlinear optical effects that provide the flexible base of the so-called virtual electrodes.

In this thesis, Michael Esseling explores for the first time photorefractive crystals as the substrate material for opto-electronic tweezers in order to realize optically patterned virtual electrodes. Apart from very high electric fields, which provide exceptionally high forces, this material class benefits from its nonlinear change in the refractive index. Michael introduces digital holography as the method-of-choice for the two-dimensional inspection of the induced electric field structure. Subsequently, he applies the knowledge about the optimal tuning conditions for dielectrophoresis to manipulate microscopic matter. He develops a method to create complex, flexible virtual electrodes, and uses them to trap and manipulate not only solid absorbing particles, but also liquid matter as droplets. Michael succeeds in realizing droplet routers as well as charge sensors by combining advanced polymer fabrication techniques with flexible optically induced dielectrophoretic manipulation, and thereby advances optical trapping in Lab-on-a-Chip devices. Michael's thesis is not only concerned with the optical and electrical properties of photorefractive optoelectronic tweezers and the solvent media under investigation, but also the reader will benefit from the detailed description of the computation and fabrication protocols employed in this thesis, e.g., a modified algorithm to significantly speed-up the phase-unwrapping process or custom-made SU8 master molds for microfluidic mixers or droplet generators. These protocols are described in a way so that both experts and newcomers can benefit from their reading.

Münster, June 2014 Prof. Cornelia Denz

Acknowledgments

As with every Ph.D. thesis, there are a lot of people who contributed to its success: First of all, I would like to express my gratitude to Prof. Cornelia Denz, not only for her continuous advice, and for supporting the experiments of this thesis and a number of promising side projects, but also for giving me the opportunity to present and discuss my results at distinguished international conferences.

Furthermore, I would like to personally thank Jörg, Patrick, Wolfgang, and Mike for extensive discussions on scientific problems and technical details, which helped me a lot. A big thank you goes to the students who contributed nice ideas and a lot of manpower to the experiments, namely Stefan Gläsener, Hannes Futterlieb, and Daniel Chaparro González, just as well as to our cooperation partners, especially to Annamaria Zaltron for preparing an almost infinite number of nice crystal samples according to the experimental needs and for her thorough proofreading of the manuscript. One of the greatest things during my time in the Nonlinear Photonics workgroup was the excellent working atmosphere and companionship I experienced. Therefore, I would like thank all present and past group members, and especially my fellow Ph.D. students for a wonderful time in the group with lots of nice activities that cheer you up even and especially in times when the lab work is not going too well. Keep it up like this!

Above all, I would like to thank my parents for supporting me, first of all morally, but let us not forget financially, so that I could do my undergraduate studies without worrying more than I do anyway. Last, but not the least, and maybe most importantly, I would like to thank Kathrin who always supported me and still puts up with me after so many years. Thank you for being you!

Contents

Chapter 1
Introduction: Optically-Mediated Particle Manipulation with High Throughput

One of the most prominent features of the human species is arguably the ever-persisting fever to breach existing limits. This urge has driven technological devices to the high degree of sophistication and integration that today dominates our live. With the advent of the optical systems such as the looking glass or the microscope, for example clockmakers were able to construct and assemble extremely fine pieces such as millimeter gears. However, if only the hand would be available in the microscopic toolset, the journey from big to small would have been over a long time ago, since ultimately each procedure would be limited by the accurateness and trembling of human hands. In the neverending quest for smaller and more accurate handling methods, optical micro manipulation techniques have been developed with the common aim to facilitate reconfigurable and precise microscopic manipulation of matter.

The suggestion that light can exert forces on matter has already been made by the German astrophysicist Kepler in the 17th century, long before mankind was aware of the electrodynamic nature of light. Kepler's postulation of optical forces was based on his observation that the dust tail of comets always points away from the sun (see [1] and references therein). A theoretical background to this observation was given when James Clerk Maxwell formulated his famous *Maxwell's equations* and deduced that electromagnetic waves must exert pressure on dielectric surfaces. Maxwell—without being able to observe his prognosis directly—already suspected that focused rays of light, incident on a surface suspended in vacuum "might perhaps produce an observable mechanical effect" [2]. It was not before the experiments of Lebedev, Crookes and Ehrenhaft that this effect was observed in the laboratory. Lebedev was able to measure the radiation pressure on absorbing and reflecting plates consisting of different materials and found his results in good agreement with Maxwell's calculations [1]. Ehrenhaft conducted similar experiments on the influence of light on microscopic particles. He observed that for some substances, the trajectory upon falling through a focused beam of light was altered either in or against the direction of light propagation. Consequently, Ehrenhaft divided the investigated substances into the classes *light-positive* and *light-negative* [3]. Neat as these observations may have been, the fascinating capabilities of all-optical micro manipulation could really

M. Esseling, *Photorefractive Optoelectronic Tweezers and Their Applications*,
Springer Theses, DOI 10.1007/978-3-319-09318-5_1

be harnessed only after the invention of the laser. In a pioneering work, Arthur Ashkin and his coworkers showed for the first time the controlled guiding and confinement of microscopic objects in a tight laser focus [4–6]. Their observations marked the birth of a novel tool that was later termed *optical tweezers*. This term was chosen due to the light's ability that—if focused strong enough—it can hold matter in three dimensions, much like mechanical tweezers. Starting from this groundbreaking discovery, the concept of using light to influence matter has been expanded into several directions to be able to fulfill different purposes. A general classification of the effects that are exploited can be made into *direct* and *indirect* interactions. In this context, *direct* denotes all interactions of photons or light fields with the particle, like refraction, polarization or transfer of momentum, whereas *indirect* describes optically-mediated manipulation, which means that a secondary thermal or electric effect, catalyzed by a structured illumination, provides the force for micromanipulation [7].

The first category of direct optical effects can be intuitively understood in the Ray optics description: The conservation of momentum requires that the refraction of light, i.e. the change of momentum of a photon, must in turn exert a force on a dielectric particle [4]. Although tiny in absolute values, this force can have a significant effect on small particles in the focus of a laser beam, where the forces from a multitude of photons add up. Stable three-dimensional confinement is typically achieved using high-numerical aperture microscope objectives [6]. In order to manipulate more than one particle at a time and move it in the focal volume, different solutions have been conceived that can adjust the laser focus in two dimensions using steerable mirrors [8] or acousto-optic deflectors [9]. The most versatile method of the beam shaping, however, is the use of a spatial light modulator (SLM) that can manipulate the phase of a light field pixel-wise. Through the replication of blazed grating and Frensel lenses, these devices – often called holographic optical tweezers (HOT)—can create many foci and manipulate them in all three dimensions without the need for mechanically moving parts [10]. For an edited collection of milestone papers about optical tweezers, the interested reader is referred to Ref. [11]. Optical tweezers have arguably become one of the most-frequently used tools in biological and material science applications [12], however, as very often in life, there is no *one-size-fits-all* solution, and there are also some drawbacks of optical tweezers. First of all, the previously described direct manipulation methods account for the optical properties of matter, so they are limited when the absorption of particles becomes too strong [13]. Whenever a photon is absorbed by a particle, the transfer of the photon momentum pushes it a little bit out of the focal position. For a particular absorption, these scattering forces, which also originate from Fresnel reflections on the boundaries, cannot be counterbalanced by the gradient forces, which drag a particle into the region of high light intensity. Over and above, even if a particle is perfectly transparent, which is quite a strong assumption in the first place, the presented schemes are limited when it comes to the manipulation of a large number of particles. In the case of holographic optical tweezers, the light power available must be divided amongst all trapping sites. Even if the laser is strong enough for a multitude of trapping sites, the SLM may still be the bottleneck in this configuration due to the limited maximum power it can accept before thermal breakdown [14].

In order to increase the throughput of optically-induced forces, researchers took one step back and took a deeper look into the existing micro manipulation techniques that already provided a high throughput. These *secondary* or *indirect* effects, in which the force is not provided by the photons themselves but rather by thermal or electrical effects, which are mediated by light, are able to provide forces orders of magnitude larger than the ones in direct optical tweezers [15]. The effect of light-induced thermophoresis, also called photophoresis, was the first light-induced secondary effect ever to be observed (in 1874 by Crookes, even earlier than Lebedev). His famous radiometer—often referred to as a *light mill*—remains one of today's most famous teaching experiments [16]. In contrast to the experiments by Lebedev, the origin of the forces here is not the change of momentum of photons, but rather the heating effect caused by their absorption. The effect of optically-mediated thermal heating is also able to explain the counter-intuitive movement of moderately absorbing matter towards the light source as observed by Ehrenhaft in 1918 [3, 17]. Photophoresis in general exploits the fact that an inhomogeneously illuminated absorbing particle transfers the heat generated by this absorption to its surrounding. The ensuing expansion of the medium causes a photophoretic force that acts away from high light intensity regions for highly absorbing particles, which become hotter on the side of incidence—denoted as *positive photophoresis*—and towards high intensity regions for moderately absorbing particles with a higher refractive index than their surrounding, where the light is focused to and which become hotter on their rear side—also known as *negative photophoresis*. By this effect, a multitude of interesting applications could be demonstrated, such as meter-scale transport of absorbing beads [18], the construction of holographic bottle traps [19] or even the manipulation of absorbing fluid droplets [20].

The other secondary effect, which forms the basis of this thesis, is named *dielectrophoresis* and relies on the electrical properties of matter rather than the optical ones. Since the 1950s, when this effect has first been described by Pohl [21], it has been developed into a powerful tool when it comes to massive sorting and actuation of dielectric particles. Dielectrophoresis (DEP) describes the forces present onto polarizable matter in an electric field. In this case, the microscopic particle can be illustratively depicted as a dipole. Naturally, the electrophoretic forces on both sides of the handle neutralize each other when the field is homogeneous. However, if the field is highly modulated, two sides of the diople experience different force magnitudes and hence a net force. In contrast to all previous methods, which all rely on the optical and thermal attributes of matter, DEP only takes into account the electrical properties. It can therefore serve as a very mighty additional tool for applications, where for example the absorption is too high for direct optical tweezing, but too low for efficient photophoresis. Particularly pleasing about this effect is the fact that it acts also on neutral polarizable matter, which means that no pre-treatment such as charging is needed. In general, all materials are more less polarizable, described by their complex permittivity. This facilitates a separation of particles according to their dielectric properties and a couple of devices have been envisaged and realized for cell transportation [22], sorting of beads according to their size or electrical properties [23, 24] and as a means to pump liquids and generate droplets [25].

The gap between the flexible optical methods, such as optical tweezers, and the high throughput of electrokinetic effects, as observed in traditional DEP devices, is bridged by the invention of so-called *virtual electrodes* for *optoelectronic tweezers* [26]. In optoelectronic tweezers, a sandwich of a liquid medium and a photoconductive layer between indium-tin-oxide (ITO) electrodes is biased by an external voltage supply. In the non-illuminated case, the voltage drops mainly over the photoconductive layer due to its very high resistance. If this layer is illuminated however, it becomes locally conductive, thereby supplying optically-switchable virtual electrodes for the external bias voltage. The inhomogeneous electric fields that are created by these electrodes can be used to manipulate dielectric matter, which greatly increased the flexibility of DEP towards optically-induced changes in the electrode structure. The invention of this new technique has triggered a number of novel concepts and geometries for the dielectrophoretic manipulation of matter. The interested reader is referred to the excellent review of Wu [27]. In order to become even more flexible and to economize the external voltage supply, Eggert et al. have used photorefractive lithium niobate ($LiNbO_3$) to trap matter on its surface [28]. $LiNbO_3$ is a photosensitive material that is able to generate very high internal electric fields solely upon inhomogeneous illumination by a special effect called the bulk photovoltaic effect. This first demonstration of optically-mediated massive particle manipulation has triggered a great interest in the use of photorefractive materials for this purpose.

This thesis presents the first investigation of *photorefractive optoelectronic tweezers* (POT) and their properties. Photorefractive materials of different constitution are used for the electrokinetic manipulation of solid and liquid matter on their surfaces. A very prominent feature that separates photorefractive materials from other photoconductors is the fact that the refractive index is changed with the applied external or internal electric field. Therefore, these materials provide optical accessibility of electric charge distributions, which makes them nicely suited to study principal phenomena, which are important also for other material classes. Optical methods, especially digital holographic microscopy (DHM) will be used to quantify the optically-created internal electric fields. Despite the fact that some publications have already demonstrated the use of $LiNbO_3$ crystals for particle trapping [29–31], a number of open questions still remain, such as the selection of an optimal substrate material and the tuning of its figures-of-merit. Additionally, there are ambivalent theoretical models about the induced polarizations and hence the equilibrium positions for particles, that are trapped on the surface of such a crystal [32]. Since the technique of DHM is able to calculate the spatial structure of internal fields in the image domain, a direct correlation between trapped matter and electric field position and form can be made. This allows a critical discussion of the resulting trap patterns for particles surrounded by different suspension media. A different point of discussion is the often-claimed but never demonstrated ability of POT to be used also in biological applications. This thesis will present an equivalent circuit model for the charge separation near the surface of the substrate material and discuss the electrical properties the solvent media for POT should possess. Regarding the orientation of the crystal, it will be investigated how the unidirectional charge transport in most photorefractive materials influences the variety of light patterns that can be used.

Last but not least, this thesis puts a great focus on the possible applications of electrokinetic effects for optofluidics. The previous publications in this field have been primarily concerned with the confinement of solid matter to a surface, however, POT can be used for many more applications. Their optical reconfigurability and good chemical resistance makes them very good active substrates for optofluidic applications, such as the sealing of polymer micro channels, in which particles can be manipulated directly out of the flow. Furthermore, the polar nature of the charge transport in $LiNbO_3$ can be exploited to construct a charge sensor or an optically-addressable trap for charged matter that is able to realize truly arbitrary particle patterns. Above all, it will be investigated how dielectrophoretic forces from POT can influence fluids, such as droplets of immiscible fluids or polymers that are liquid in their uncured state.

References

1. P. Lebedev, Experimental examination of light pressure. Ann. der Phys. **6**, 433 (1901)
2. J. Maxwell, *A treatise on Electricity and Magnetism*, vol. 2. (Clarendon Press, Oxford, 1873)
3. F. Ehrenhaft, Die Photophorese. Ann. der Phys. **4**(56), 81–132 (1918)
4. A. Ashkin, Acceleration and trapping of particles by radiation pressure. Phys. Rev. Lett. **24**(4), 156 (1970)
5. A. Ashkin, J. Dziedzic, Optical levitation by radiation pressure. Appl. Phys. Lett. **19**, 283 (1971)
6. A. Ashkin, J.M. Dziedzic, J.E. Bjorkholm, S. Chu, Observation of a single-beam gradient force optical trap for dielectric particles. Opt. Lett. **11**, 288–290 (1986)
7. M. Woerdemann, C. Alpmann, M. Esseling, C. Denz, Advanced optical trapping by complex beam shaping. Laser Photonics Rev. **7**(6), 839–854 (2013)
8. C. Mio, T. Gong, A. Terray, D. Marr, Design of a scanning laser optical trap for multiparticle manipulation. Rev. Sci. Instrum. **71**(5), 2196–2200 (2000)
9. K. Vermeulen, J. van Mameren, G. Stienen, E. Peterman et al., Calibrating bead displacements in optical tweezers using acousto-optic deflectors. Rev. Sci. Instrum. **77**(1), 013704 (2006)
10. J.E. Curtis, B.A. Koss, D.G. Grier, Dynamic holographic optical tweezers. Opt. Commun. **207**, 169–175 (2002)
11. M.J. Padgett, J.E. Molloy, D. McGloin (eds.), *Optical Tweezers—Methods and Applications* (CRC Press, 2010)
12. F. Hoerner, M. Woerdemann, S. Mueller, B. Maier et al., Full 3D translational and rotational optical control of multiple rod-shaped bacteria. J. Biophotonics **3**(7), 468–475 (2010)
13. H. Rubinsztein-Dunlop, T.A. Nieminen, M.E.J. Friese, N.R. Heckenberg, *Advances in Quantum Chemistry: Modern Trends in Atomic Physics, Chapter Optical Trapping of Absorbing Particles*, (Academic Press, San Diego, 1998), pp. 469–492
14. M. Woerdemann, F. Holtmann, C. Denz, Holographic phase contrast for dynamic multiple-beam optical tweezers. J. Opt. A Pure Appl. Opt. **11**(3), 034010 (2009)
15. V.G. Shvedov, A.S. Desyatnikov, A.V. Rode, W. Krolikowski et al., Optical guiding of absorbing nanoclusters in air. Opt. Express **17**(7), 5743–5757 (2009)
16. W. Crookes, On attraction and repulsion resulting from radiation. Philos. Trans. Roy. Soc. Lond. **164**, 501–527 (1874)
17. S. de Angelis, A. Finizo, P. Mormile, G. Pierattini et al., Experimental results on the photophoretic motion and radiometric trapping of particles by irradiation with laser-light. Appl. Phys. B Photophys. Laser Chem. **47**(3), 247–250 (1988)

18. V.G. Shvedov, A.V. Rode, Y.V. Izdebskaya, A.S. Desyatnikov et al., Giant Optical Manipulation. Phys. Rev. Lett. **105**(11), 118103 (2010)
19. C. Alpmann, M. Esseling, P. Rose, C. Denz, Holographic optical bottle beams. Appl. Phys. Lett. **100**, 111101 (2012)
20. M. Esseling, P. Rose, C. Alpmann, C. Denz, Photophoretic trampoline-Interaction of single airborne absorbing droplets with light. Appl. Phys. Lett. **101**(13), 131115 (2012)
21. H. Pohl, Some effects of nonuniform fields on dielectrics. J. Appl. Phys. **29**, 1182–1188 (1958)
22. C.H. Kua, Y.C. Lam, I. Rodriguez, C. Yang et al., Dynamic cell Fractionation and transportation using moving dielectrophoresis. Anal. Chem. **79**(18), 6975–6987 (2007)
23. J. Rousselet, G. Markx, R. Pethig, Separation of erythrocytes and latex beads by dielectrophoretic levitation and hyperlayer field-flow fractionation. Colloids Surf. A Physicochem. Eng. Aspects **140**(1–3), 209–216 (1998)
24. P. Gascoyne, J. Vykoukal, Particle separation by dielectrophoresis. Electrophoresis **23**(13), 1973–1983 (2002)
25. T. Jones, M. Gunji, M. Washizu, M. Feldman, Dielectrophoretic liquid actuation and nanodroplet formation. J. Appl. Phys. **89**(2), 1441–1448 (2001)
26. P. Chiou, A. Ohta, M. Wu, Massively parallel manipulation of single cells and microparticles using optical images. Nature **436**, 370–372 (2005)
27. M.C. Wu, Optoelectronic tweezers. Nat. Photonics **5**(6), 322–324 (2011)
28. H.A. Eggert, F.Y. Kuhnert, K. Buse, J.R. Adleman et al., Trapping of dielectric particles with light-induced space-charge fields. Appl. Phys. Lett. **90**, 241909 (2007)
29. S. Grilli, P. Ferraro, Dielectrophoretic trapping of suspended particles by selective pyroelectric effect in lithium niobate crystals. Appl. Phys. Lett. **92**, 232902 (2008)
30. X. Zhang, J. Wang, B. Tang, X. Tan et al., Optical trapping and manipulation of metallic micro/nanoparticles via photorefractive crystals. Opt. Express **17**, 9981–9988 (2009)
31. M. Esseling, F. Holtmann, M. Woerdemann, C. Denz, Two-dimensional dielectrophoretic particle trapping in a hybrid crystal/PDMS-system. Opt. Express **18**(16), 17404–17411 (2010)
32. J. Villarroel, H. Burgos, A. Garcia-Cabanes, M. Carrascosa et al., Photovoltaic versus optical tweezers. Opt. Express **19**(24), 24320–24330 (2011)

Chapter 2
Electrokinetic Forces in Inhomogeneous Fields

This second chapter will describe the fundamentals of electrokinetic forces on particles suspended in a solvent medium. It starts from the basic observation of electrophoretic motion, which occurs for charged particles and summarizes its main properties. The main focus is on the derivation of a force equation for uncharged but polarizable particles that are subject to a strongly inhomogeneous electric field. In this field, the internal polarization of a micro-object interacts with the field, an effect called *dielectrophoresis*, which forms the basis for particle manipulation in this thesis.

2.1 Electrophoresis and Dielectrophoresis

For particles suspended in an electric field, the first observation in daily life is that a particle of charge q is influenced by other charges and the resulting electric field E:

$$F_{\mathrm{EP}} = q \cdot E. \tag{2.1}$$

This effect is termed *electrophoresis*, and it is a standard school experiment for electrostatic interaction to rub a cotton cloth over a glass bar to separate charges on both parts and to demonstrate their mutual attraction. Furthermore, it can be employed in more useful ways to construct lenses in electron microscopy [1] or to manipulate charged matter in an electric field [2]. The predominant properties of electrophoretic movement can be described as follows:

- Particles must be charged to be affected by an electric field.
- Electrophoresis occurs regardless of the spatial structure of the electric field.
- A reverse of the sign of the charge or the orientation of the electric field is accompanied by a reverse of particle motion.
- The mean displacement of electrophoretic motion vanishes in an alternating electric field.

M. Esseling, *Photorefractive Optoelectronic Tweezers and Their Applications*,
Springer Theses, DOI 10.1007/978-3-319-09318-5_2

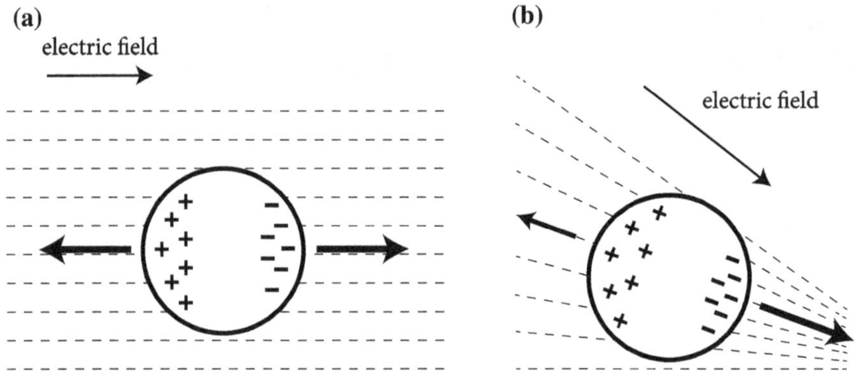

Fig. 2.1 Polarizable particle influenced by an electric field: in **a** the particle is polarized, but the forces on both side counteract, while in **b** a *positive dielectrophoretic force* moves particle towards regions of high field intensity

The charging of particles, while being easy for macroscopic objects, may not be easily accomplished for microscopic objects. Once the micro particle comes into contact with the electrode of opposite charge, electrons may be transferred and the interaction is lost.

Another effect that is able to manipulate matter on the micro scale without the presence of excess charges was first described in 1958 by Pohl under the term *dielectrophoresis* [3], which is an extension of the term "electrophoresis" to include also dielectric, neutral particles. Dielectrophoresis (DEP) occurs when polarizable matter is subject to an inhomogeneous electric field. A descriptive illustration can be found in Fig. 2.1. If a polarizable particle is placed in a homogeneous electric field, free charge carriers on the particle become separated, which means that a dipole is induced. This separation is strongly dependent on the electric properties of the particle and the surrounding medium, as will be seen later. This dipole is influenced by counteracting forces of the same magnitude on both sides, hence it experiences no net force. A different situation occurs when the external electric field is inhomogeneous (cf. Fig. 2.1b). In this case the dipole experiences a force of higher magnitude on the side where the field intensity is higher, which causes the particle to be attracted to this region. This behavior is referred to as *positive dielectrophoresis* for reasons described later in this chapter. Of course this simple image neglects several effects that are important in the description of dielectrophoresis. The induced electrodynamic moment may not always be a dipole but consist of quadrupole or even more complex multipolar moments [4, 5]. Additionally, the electric field does not necessarily have to be constant, but can change in magnitude and even sign; the force may not always be attractive towards the regions of high field intensity but could as well be repulsive. To come to a more accurate description of the forces, the *effective moment method* as suggested by Jones [4] will be described in the following.

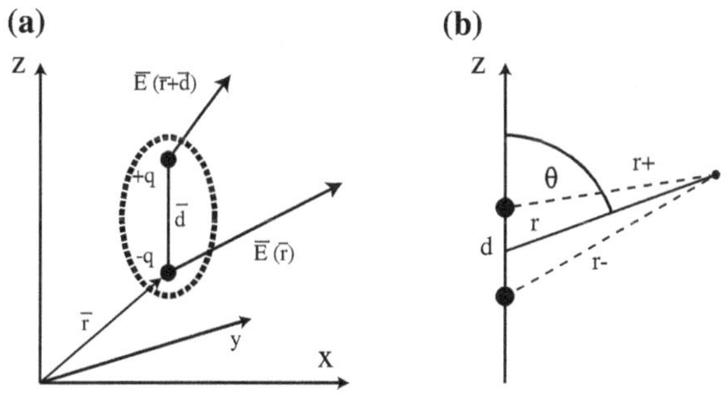

Fig. 2.2 Illustrative scheme for the calculation of the dielectrophoretic force, **a** net force on a dipole, **b** effective moment calculation

2.2 Dielectrophoretic Force Calculation

The calculation of the dielectrophoretic force on a microscopic particle suspended in a medium has to account for the polarization of a particle and its ensuing interaction with the electric field. An intriguing first approach is to consider a point dipole that is subject to a non-uniform field. In this case the net force can be described as the difference between the two electrophoretic forces acting on the two ends of the dipole [4]:

$$\vec{F} = \vec{F}_{\mathrm{EP}}(+q) + \vec{F}_{\mathrm{EP}}(-q) = q\vec{E}(\vec{r} + \vec{d}) - q\vec{E}(\vec{r}). \tag{2.2}$$

Using a Taylor expansion and introducing ∇ as the gradient operator for the electric field, the force can be approximated as

$$\vec{F} = \vec{p}\nabla\vec{E} + \ldots, \tag{2.3}$$

where higher order terms have been neglected and $\vec{p} = q\vec{d}$ has been introduced as the dipole moment of a particle of size d. The challenge in describing the force is the calculation of the dipole moment for microscopic polarizable particles and to relate this moment to the particle properties. Jones suggested to use an effective moment method to derive the dipole moment induced inside a dielectric particle [6]. The effective moment technique calculates the dipolar potential for a infinitesimal dipole suspended in an isotropic liquid. This potential is compared to the solution of Laplace's equation for a microscopic particle suspended in the same liquid, but under the influence of a homogeneous electric field, which induces a very similar dipole structure inside the dielectric particle.

From Fig. 2.2b, it is obvious that the potential Φ of a dipole can be calculated as the superposition of two charge potentials separated by a distance d [6]:

$$\Phi(r, \theta) = \frac{q}{4\pi\varepsilon_1 r_+} - \frac{q}{4\pi\varepsilon_1 r_-}, \tag{2.4}$$

where ε_1 is the static permeability of the medium in which the charges are located. Using the geometric relation $r_\pm = (1 + (\frac{d}{2r})^2 \mp \frac{d}{r}\cos\theta)^{-1/2}$ and expanding this using its Taylor series, the potential of a dipole of finite extent can be written as [4]

$$\Phi_{\text{dipole}} = \frac{qd\cos\theta}{4\pi\varepsilon_1 r^2} + \frac{qd^3(5\cos^3\theta - 3\cos\theta)}{32\pi\varepsilon_1 r^4} + \ldots, \tag{2.5}$$

where the first term describes the potential of an infinitely small dipole and the second term is an octupole correction due to the finite extent of the charge separation. In order to relate the effective dipole moment in Eq. 2.5 to the particle properties, it is necessary to solve Laplace's equation with appropriate boundary conditions for a homogeneous sphere of radius R suspended in a constant electric field. In this case, the radial symmetric electrostatic potential can be written as the superposition of the dipole potential and the external electric field E_0 [4]:

$$\begin{aligned} \Phi_1(r,\theta) &= E_0 r\cos\theta && \text{for r > R} \\ \Phi_2(r,\theta) &= E_0 r\cos\theta + \frac{A\cos\theta}{r^2} && \text{for r} \leq \text{R}. \end{aligned} \tag{2.6}$$

It is required that both the potentials and the electromagnetic displacement flux across the boundary of sphere and fluid are continuous. With these assumptions, the unknown constant A can be determined to be [4]

$$A = \frac{\varepsilon_p - \varepsilon_m}{\varepsilon_p + 2\varepsilon_m} R^3 E_0, \tag{2.7}$$

where $\varepsilon_{p,m}$ are the relative dielectric permittivities of the sphere (particle) and the surrounding medium, respectively.

Comparing this to the potential of a point dipole in Eq. 2.5, an expression can be derived for the effective dipole moment p_{eff}

$$p_{\text{eff}} = 4\pi\varepsilon_m R^3 E_0 \frac{\varepsilon_p - \varepsilon_m}{\varepsilon_p + 2\varepsilon_m}. \tag{2.8}$$

Inserting this expression for the effective dipole moment into Eq. 2.3 and making use of the relationship $E_0 \nabla E_0 = \frac{1}{2}\nabla E_0^2$, one arrives at the often quoted expression for the dielectrophoretic force on a homogeneous sphere in a dielectric medium [3, 6, 7]:

$$F_{\text{DEP}} = 2\pi R^3 \varepsilon_1 K(\varepsilon_m, \varepsilon_p) \nabla E_0^2, \tag{2.9}$$

where $K(\varepsilon_m, \varepsilon_p)$ is the material- and shape-dependent Clausius-Mossotti factor that determines the magnitude and direction of the dielectrophoretic force. Note that this derivation of the force is only valid for a perfectly insulating sphere in a dielectric, non-conducting liquid. Using the same approach as before with minor extensions, a finite conductivity and alternating electric fields of frequency ω can be included to arrive at a more general form of the force [6]:

$$\langle F(\omega)_{\mathrm{DEP}}\rangle = 2\pi R^3 \varepsilon_m \Re(K(\omega))\nabla \left\langle E_{\mathrm{rms}}^2 \right\rangle, \tag{2.10}$$

where $\langle \ldots \rangle$ defines the temporal average, $\Re$ the real part and E_{rms} the root mean square amplitude of the alternating electric field. For a particle of spherical shape, the Clausius-Mossotti factor is defined as [3]

$$K(\omega) = \frac{\varepsilon_p^* - \varepsilon_m^*}{\varepsilon_p^* + 2\varepsilon_m^*}, \tag{2.11}$$

with ε_k^* as the complex permittivity of the medium and sphere, respectively:

$$\varepsilon_k^* = \varepsilon_k + i\frac{\sigma_k}{\omega}. \tag{2.12}$$

2.3 Clausius-Mossotti Factor

Based on the electric properties of the materials under investigation, the real part of the complex Clausius-Mossotti factor determines the magnitude and direction of the dielectrophoretic force. If the real part is positive, the suspended sphere is attracted towards regions of high field intensity, commonly referred to as *positive dielectrophoresis*. The opposite case of *negative dielectrophoresis* occurs if the real part is negative and matter is repelled by high field intensities. By looking at Eq. 2.11, it becomes obvious that regardless of the electric properties of the materials under investigation, the real part of the Clausius-Mossotti factor for a spherical particle is fixed between $\frac{1}{2} \leq K(\omega) \leq 1$. It is important to note that this restriction does not apply in all cases. For example, elongated structures like nanowires can possess values of $K(\omega)$ exceeding 1 by orders of magnitude [8], accompanied by substantially higher dielectrophoretic forces. Reference [4] gives further analytical expressions for the Clausius-Mossotti factor in the case of elliptical particles where the polarization along the longer and shorter axes are different and torques are exerted onto the structures. Even more complex forms, for which there is no analytical solution for the induced multipolar moments, have been treated numerically [9].

For many objects, however, the assumptions of a spherical shape is a good approximation and allows to predict dielectrophoretic forces. If objects under investigation deviate strongly from this shape, the force calculation has to be changed accordingly. Looking at Eqs. 2.11 and 2.12, it is obvious that the Clausius-Mossotti factor has two limiting values of

$$K(\omega \to 0) = \frac{\sigma_p - \sigma_m}{\sigma_p + 2\sigma_m} \tag{2.13}$$

for low frequencies, which means that in the case of slowly varying alternating fields the Clausius-Mossotti factor can be calculated solely knowing the respective conductivities, whereas for very high frequencies only the dielectric permittivities govern its magnitude:

$$K(\omega \to \infty) = \frac{\varepsilon_p - \varepsilon_m}{\varepsilon_p + 2\varepsilon_m}. \tag{2.14}$$

In between these limiting cases, the Clausius-Mossotti factor must be calculated taking into account both conductivity and permittivity. In particular, it can change its sign when the frequency is changed, for example, if a medium and particle are used where $\varepsilon_2 > \varepsilon_1$, but $\sigma_2 < \sigma_1$. Then for low frequencies, a suspended particle is repelled by field gradients into regions of low field intensity, whereas for high frequencies, it is attracted into high intensity areas. Analogous to the phenomenon of electrophoresis, the most important properties of dielectrophoresis can be summarized as follows:

- Dielectrophoresis acts on polarizable matter, even when it is uncharged.
- It occurs only in inhomogeneous electric fields.
- Due to the fact that the force is proportional to the gradient of E^2, its direction does not change upon repoling the electric field.
- Depending on the sign and magnitude of the Clausius-Mossotti factor, suspended particles are either attracted towards or repelled from regions of high field intensity.

Through these properties, DEP has lended itself to a multitude of sorting and manipulation concepts. These concepts either exploit that for suitable combinations of two different particles in a medium, the sign of the Clausius-Mossotti factor can be different [10], which leads to the spatial separation of particles or they aim at the strong size dependence of the force, which strongly affects the height to which particles can be levitated. This behavior enables a different speed of transport in pressure-driven Hagen-Poiseuille flows, a technique commonly denoted as flow-field fractionation [11, 12]. It is an interesting but often neglected fact about dielectrophoresis and optical tweezers, which were mentioned in the history of optically-induced particle trapping, that—while both tools are often described as two separate phenomena—both forces originate from the same description in the approximation of particles smaller than the electric field inhomogenity and that both can be reduced to the same form. It has been previously mentioned in Eq. 2.14 that in the limit of very high frequencies the Clausius-Mossotti factor is calculated by the relative permittivities. Considering that the refractive index for non-magnetic matter ($\mu_r = 1$) is defined as $n = \sqrt{\varepsilon_r}$, Eq. 2.14 becomes:

$$K(\omega \to \infty) = \frac{n_p^2 - n_m^2}{n_p^2 + 2n_m^2}, \tag{2.15}$$

which is nothing else than the polarizability α that is typically used in the field of optical tweezers to describe the magnitude and direction of optical forces in the so-called Rayleigh regime [13, 14], which corresponds to the case of subwavelength particles. It should be noted however that the term *refractive index* is typically only used for the electromagnetic spectrum from UV to infrared, while in DEP literature the electric properties are summarized in the term *complex permittivity*, as described before.

2.4 Generalization of DEP for Large Objects and Continuous Media: Multipoles and Polarization Force Density

So far the calculation of the dielectrophoretic forces has only considered dipolar moments, which is valid if the field inhomogeneity is much larger than the particle itself. However, for the sake of completeness, it should be mentioned that situations may arise where this approximation does not hold any more, for example if the particles size is comparable to the electrode structure size or if the object under investigation is not a discrete sphere any more, but rather a continuous medium. The cases of a large sphere or any other object can be described by the general theory of multipoles of order n [5]:

$$\vec{\vec{\vec{p}}}^{(n)} = \frac{4\pi\varepsilon_m R^{2n+1}}{(2n-1)!!} K^{(n)} (\nabla)^{n-1} \vec{E}, \tag{2.16}$$

with the higher order Clausius-Mossotti factor

$$K^{(n)} = \frac{\varepsilon_p^* - \varepsilon_m^*}{n\varepsilon_p^* + (n+1)\varepsilon_m^*}. \tag{2.17}$$

In the case of non-conducting dielectric fluids, a generalization of the force is made by introducing the *Kelvin polarization force density* $\vec{P}$, where the molecules of the fluid are considered as infinitesimal dipoles. $\vec{P}$ is equivalent to density of the dipoles N_p, multiplied by their dipole value $\vec{p}$ and the total force is calculated as the integral over the continuous volume that is to be actuated [15]:

$$\vec{F} = \int (N_P \vec{p})\, \nabla \vec{E} dV = \frac{1}{2}\varepsilon_0 \int (\varepsilon_l - \varepsilon_m) \nabla \vec{E}^2 dV, \tag{2.18}$$

where $(\varepsilon_l - \varepsilon_m)$ describes the *excess polarization* of a dielectric substance l surrounded by a medium m [16]. Note that this derivation treats each dipole as independent and neglects the mutual interaction of different dipoles. Nevertheless, the Kelvin polarization density can be used to qualitatively explain the behavior of liquids later in this thesis. In general, it states that the same assumptions as before also apply to continuous media, namely that a medium is attracted to high field intensity regions in the case of a positive force, i.e. when it possesses a higher polarizability than its surrounding, and repelled if the total force is negative.

References

1. J. Orloff, L. Swanson, Asymmetric electrostatic lenses for field-emission microprobe applications. J. Appl. Phys. **50**(4), 2494–2501 (1979)
2. A. Cohen, W. Moerner, Method for trapping and manipulating nanoscale objects in solution. Appl. Phys. Lett. **86**(9), 093109 (2005)
3. H. Pohl, Some effects of nonuniform fields on dielectrics. J. Appl. Phys. **29**, 1182–1188 (1958)
4. T. Jones, *Electromechanics of Particles.* (Cambridge University Press, Cambridge, 1995)
5. T. Jones, M. Washizu, Multipolar dielectrophoretic and electrorotation theory. J. Electrost. **37**(1–2), 121–134 (1996)
6. T. Jones, Basic theory of dielectrophoresis and electrorotation. IEEE Eng. Med. Biol. Mag. **22**(6), 33–42 (2003)
7. H.A. Eggert, F.Y. Kuhnert, K. Buse, J.R. Adleman et al., Trapping of dielectric particles with light-induced space-charge fields. Appl. Phys. Lett. **90**, 241909 (2007)
8. A. Jamshidi, P.J. Pauzauskie, P.J. Schuck, A.T. Ohta et al., Dynamic manipulation and separation of individual semiconducting and metallic nanowires. Nat. Photonics **2**(2), 85–89 (2008)
9. N.G. Green, T.B. Jones, Numerical determination of the effective moments of non-spherical particles. J. Phys. D Appl. Phys. **40**(1), 78–85 (2007)
10. H. Morgan, M. Hughes, N. Green, Separation of submicron bioparticles by dielectrophoresis. Biophys. J. **77**(1), 516–525 (1999)
11. J. Rousselet, G. Markx, R. Pethig, Separation of erythrocytes and latex beads by dielectrophoretic levitation and hyperlayer field-flow fractionation. Coll. Surf. A Physicochem. Eng. Asp. **140**(1–3), 209–216 (1998)
12. H.-S. Moon, K. Kwon, S.-I. Kim, H. Han et al., Continuous separation of breast cancer cells from blood samples using multi-orifice flow fractionation (MOFF) and dielectrophoresis (DEP). Lab Chip **11**(6), 1118–1125 (2011)
13. P. Smith, A. Ashkin, W. Tomlinson, 4-wave mixing in an artifical Kerr medium. Opt. Lett. **6**(6), 284–286 (1981)
14. A. Ashkin, J.M. Dziedzic, J.E. Bjorkholm, S. Chu, Observation of a single-beam gradient force optical trap for dielectric particles. Opt. Lett. **11**, 288–290 (1986)
15. H. Hauser, J. Melcher, Electromagnetic fields and energy (2013). Prentice-Hall, NJ, http://ocw.mit.edu. (1989)
16. S. Schmid, C. Hierold, A. Boisen, Modeling the Kelvin polarization force actuation of micro- and nanomechanical systems. J. Appl. Phys. **107**(5), 054510 (2010)

Chapter 3
Electric Fields and Their Detection in Photorefractive Crystals

Photorefractive crystals will be introduced in this chapter as a class of materials that can be used for the creation of *virtual electrodes*. In contrast to fixed electrodes which were used for dielectrophoresis until recently, virtual electrodes are optically induced and hence benefit from optical flexibility. This chapter will illuminate the basic properties of photorefractive crystals that are relevant not only for the optical creation of internal electric fields, but also for their visualization. It includes a description of the formation of internal space-charge fields using Kukhtarev's band transport model and then focuses on the specific generation of the internal fields using either drift in an externally applied field or the bulk photovoltaic effect. As a unique tool, the electro-optic properties of photorefractive samples can be exploited to instantly quantify the electric field two-dimensionally. Three measurement approaches that allow the determination of the internal electric field via the *linear electro-optic* or *Pockels effect* will be described and evaluated.

3.1 Optical Induction of Virtual Electrodes

In the previous chapter, it was derived how charged and uncharged particles can be manipulated using electric fields of different constitution, independent of the way these fields were created. Chapter 2 has mentioned some applications of electrokinetic manipulation, all of which were realized using fixed metal electrodes. The fabrication of these electrodes, however, remains a laborious process, since a mask has to be fabricated which can then be transferred to a metal surface via photo-lithography. Over and above, etching steps on the surface may involve hazardous substances and specialized equipment, such as acid-proof surfaces or a clean room [1]. Granted that many of these steps are nowadays daily routine and allow the fabrication of metal structures in a reasonable amount of time, the structure—once finished—is still fixed, which means that a change of the electric field distribution is impossible and requires the repetition of all process steps.

M. Esseling, *Photorefractive Optoelectronic Tweezers and Their Applications*, Springer Theses, DOI 10.1007/978-3-319-09318-5_3

Such a repetition can be economized if the electrode structure is induced all-optically, since light fields are inherently customizable by changing the phase or amplitude pattern of the light, for example with liquid crystal displays that are part of mobile phones and television sets in every households. As a connection between the previously described phenomenon of dielectrophoresis and the optical world, Chiou et al. invented what they termed *optoelectronic tweezers* [2]. This device consists of a sandwich structure of base and top layer indium tin oxide (ITO) transparent electrodes, which provide the sufficiently high voltages for efficient dielectrophoretic manipulation. The actual structure of electrodes is imprinted on a photoconductive layer by a low-power illumination pattern. Through the absorption of photons, the illuminated regions of the photoconductive layer act as so called *virtual electrodes* that can be rapidly changed by altering the illumination pattern. The change in the conductivity of this layer results in an inhomogeneous electric field near this electrode, enabling a number of possible applications for dielectrophoresis on a single device [3]. In contrast to this arrangement, in *photorefractive optoelectronic tweezers* (POT), an inhomogeneous electric field is formed through charge transport processes which act in-plane and along the surface of the crystalline material. This has the advantage over the previously described techniques that it leaves the crystal surface accessible to the application of suspensions or microchannels, hence lending itself to an easy integration into microfluidic devices [4, 5].

3.2 Photorefractive Crystals and Kukhtarev's Band Transport Model

Experimental investigations performed by Ashkin et al. in the 1960s about the non-linear properties of lithium tantalate ($LiTaO_3$) and lithium niobate ($LiNbO_3$) showed the surprising result that—even under the influence of continuous laser light of relatively low power—the refractive index of the materials was reversibly modified. The discoverers of this effect—not foreseeing the potential of their findings at first—denounced it as "highly detrimental" [6]. Shortly after this observation, first models were developed to shed light on the underlying physics [7] and it was discovered that several materials show a similar behavior [8–11] which, according to the materials' ability to change their refractive index upon non-uniform illumination, was later termed *photorefractivity* [12]. The reason for this property lies in the change of the refractive index caused by an external or internal electric field in crystals that lack inversion symmetry, also known as the *linear electro-optic* or *Pockels effect* [13]. The surprising thing about Ashkin's discovery was that the crystals under investigation did not even have an externally applied field. Some materials, among them the previously described $LiNbO_3$ and $LiTaO_3$, are able to generate high electric fields upon illumination solely by intrinsic charge separation inside the bulk, known as the *bulk photovoltaic effect* [14].

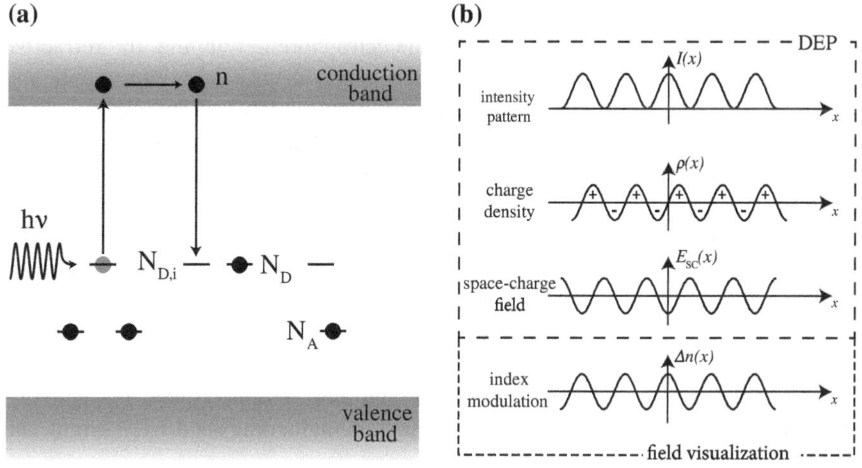

Fig. 3.1 The photorefractive effect and its use for DEP: **a** in Kukhtarev's band transport model, electrons are photo-excited into the conduction band and retrapped at ionized donor sites; **b** four-step description of photorefractive grating formation; while a modulated space-charge field is sufficient for the occurrence of DEP, the resulting refractive index modulation is useful for visualizing the electric field

To describe the formation and temporal evolution of the internal space-charge fields that are responsible for the modification of the refractive index, Kukhtarev et al. developed a band transport model similar to solid-state physics [15, 16]. In this model, visualized in Fig. 3.1, the photorefractive crystal's energy diagram is described as a semiconductor with a filled valence band and empty conduction band. In the band gap between the two bands, discrete energy levels, so-called acceptor and donor states, are present. These states are present in the photorefractive crystals either as intrinsic defects—such as vacancies or atoms on the "wrong" position in the crystal lattice, so-called *antisite defects*—or by doping the crystal with appropriate donor impurities, which can also change the spectral sensitivity of the sample [17–22]. In particular, for $LiNbO_3$, which will be used mainly in this thesis, such impurities appear as niobium anti-sites, where niobium atoms occupy lithium positions in the lattice or by dopants such as copper or iron.

Phenomenologically, the formation of internal space-charge fields and the ensuing refractive index modulation inside a photorefractive crystal can be described in four steps as visualized in Fig. 3.1b: Photons of sufficient energy can excite electrons from donor atoms to the conduction band, where they are free to travel under the influence of drift in an external electric field, diffusion or the bulk photovoltaic effect, until being trapped by an ionized donor atom. Provided that this donor atom is located in a region of high light intensity, the electron can be re-excited and travel further. The continuous repetition of this process leads to a charge redistribution from bright to dark areas until the resulting electric fields counteract the above mentioned driving mechanisms. The field created by these internal charges is called the space-charge field E_{SC} and it directly modifies the refractive index of the material via the linear electro-optic effect. Although the highly modulated internal fields alone would be sufficient for optically-induced dielectrophoretic trapping, the refractive index modulation is a very useful tool since it converts electric fields into an optically accessible quantity. Note that a number of different models and extensions of this basic model have been developed to include holes as charge carriers or more than one type of donor state [23–25]. However, even if there are different types of impurities present in the material, for low light intensities such as the ones used in this thesis ($I < 100\,\mathrm{kWm^{-2}}$), the one-center model, meaning that only one type of donor impurity is considered, is sufficient in order to explain the basic phenomena of the photorefractive effect. The following theoretical derivation follows the description of Yeh [13], who has summarized the most important features of Kukhtarev's band transport model.

Let N_D and $N_{\mathrm{D},i}$ be the density of donor and ionized donor states, respectively. Then, the generation of free electrons n in the conduction band by thermal and photonic excitation can be described as

$$\frac{\partial n}{\partial t} = (\beta + sI)(N_\mathrm{D} - N_{\mathrm{D},i}), \tag{3.1}$$

where β is the rate of thermal excitation, s is the cross-section for photoionization and I is the incident light intensity. At room temperature and for materials with a sufficiently high band gap, the thermal ionization of donor states may be neglected. Free electrons interact with previously ionized donor sites, where they are retrapped at the recombination rate γ_r. Considering all processes leads to the following rate equation for the generation of ionized donor sites [13]

$$\frac{\partial N_{\mathrm{D},i}}{\partial t} = sI(N_\mathrm{D} - N_{\mathrm{D},i}) - \gamma_\mathrm{r} n N_{\mathrm{D},i}. \tag{3.2}$$

The same rate equation applies for the excitation of electrons to the conduction band, except that the donor sites are spatially fixed, whereas the electrons are mobile and may be redistributed. The movement of electrons leads to a current density $\vec{j}$ which connects the two quantities:

$$\frac{\partial n}{\partial t} - \frac{\partial N_{\mathrm{D},i}}{\partial t} = -\frac{1}{e}\nabla\vec{j}, \tag{3.3}$$

where e is the electron charge. The current density arises due to the three different phenomena of *diffusion*, *bulk photovoltaic effect* (PVE) and *drift* in the case of an applied external field:

$$\vec{j} = \vec{j}_\mathrm{diff} + \vec{j}_\mathrm{pve} + \vec{j}_\mathrm{drift} = k_b T \mu_n \nabla n + \vec{p}I - en\mu_n\vec{E}, \tag{3.4}$$

where k_b is the Boltzmann constant, T the temperature, μ_n the electron mobility, $\vec{p}$ the photovoltaic transport vector and $\vec{E}$ the externally applied electric field. The resulting electric field can be calculated from the charge density by Gauss' law [13]:

$$\nabla \varepsilon \vec{E} = e(n + N_A - N_{D,i}), \tag{3.5}$$

with N_A as the density of the acceptor sites, which are introduced in the model to ensure charge neutrality in the absence of light. These equations, also known as *Kukhtarev's equation*, in general possess no closed form analytical solution. Typically, they are solved by restricting the solution to the steady state, i.e. setting all time-derivatives to zero, and using an ansatz of periodic light illumination, so that $I(\vec{r}) = I_0 + \Re\{I_1 e^{-i\vec{K}\vec{r}}\}$. The periodic pattern has a wave vector $\vec{K}$ with a small degree $m = I_1/I_0 \ll 1$ of modulation, which allows to use a harmonic expansion of all respective quantities $n(\vec{r})$, $N_{D,i}(\vec{r})$, $\vec{j}(\vec{r})$ and $\vec{E}(\vec{r})$ [13]. The case of a periodic sinusoidal illumination can be solved using these approximations and gives rise to a modulated internal space-charge field E_{SC}. Without any loss of generality, the interference pattern as well as the applied external field can be assumed to be along the z axis of the material, so $\vec{K} = K\vec{e}_z$, where $K = 2\pi/\Lambda$ is the k-vector of the incident light pattern with period Λ. The final solution, including contributions from all three charge transport processes, is [13]:

$$E(z,t) = E_{SC}\cos(Kz)(1 - e^{\frac{t}{\tau}}), \tag{3.6}$$

which describes a monoexponential evolution of the field strength, until the steady-state space charge field

$$E_{SC} = m \left\{ \frac{E_0 - E_{pve} - iE_d}{1 + \frac{E_d}{E_q} + i\frac{E_0 - E_{pve}}{E_q}} \right\} \tag{3.7}$$

is reached. In this equation, the respective field distributions are the externally applied field E_0, the saturation field

$$E_q = -\frac{eN_A(N_D - N_A)}{\varepsilon N_D}\frac{1}{K}, \tag{3.8}$$

which describes the highest achievable space-charge field, provided that all available trapping centers are filled, the diffusion field

$$E_d = -\frac{k_b T}{e}K, \tag{3.9}$$

created by the concentration gradient of photo-excited electrons in the conduction band, and the photovoltaic field

$$E_{pve} = -\frac{p\gamma_r N_A}{e\mu s}. \tag{3.10}$$

Note that in any case and for any field distribution, the magnitude of the internal space-charge field is ultimately limited by the saturation field due to the scaling factor in the denominator of Eq. 3.7. Furthermore, the imaginary unit of the diffusion field in Eq. 3.7 indicates that the field is phase-shifted with respect to the incident light pattern, while photovoltaic and drift field are in phase or 180° out of phase, depending on the direction of the applied field.

3.3 Internal Fields for Dielectrophoretic Trapping

In the previous description, three contributions to the internal space-charge field could be identified. Assuming that materials can be found where only one of the processes has a significant impact, we can divide those materials into the following classes:

$$\begin{aligned} E_{\mathrm{SC}} &\approx im\frac{E_{\mathrm{d}}}{1+\frac{E_{\mathrm{d}}}{E_{\mathrm{q}}}} && \text{for diffusion-dominated materials} \\ E_{\mathrm{SC}} &\approx m\frac{E_{0}}{1+\frac{E_{0}}{E_{\mathrm{q}}}} && \text{for drift-dominated materials} \\ E_{\mathrm{SC}} &\approx m\frac{E_{\mathrm{pve}}}{1+\frac{E_{\mathrm{pve}}}{E_{\mathrm{q}}}} && \text{for photovoltaic materials.} \end{aligned} \tag{3.11}$$

For drift and photovoltaic fields, the magnitude of the internal space-charge field is independent of the spatial period of the incident light pattern that creates them. In the case of the diffusion field, however, the field decays inversely with the grating period. This allows a classification of the physical mechanism that can be used to create optoelectronic tweezers on photorefractive surfaces. Let the application of POT be to capture particles of diameter $d = 2\,\mu\mathrm{m}$. In order for the dipole approximation to hold (cf. Eq. 2.5), the length scale of the electrode pattern should be much larger than the diameter of the particle. Even if we assume a spatial separation for the virtual electrodes of only twice the particle size ($K = 1.57 \times 10^6\mathrm{m}^{-1}$), the resulting diffusion space-charge field at room temperature is approximately 40 V mm^{-1}, which has only a minor influence compared to the other field contributions. The drift field, limited by the external field magnitude, may easily reach several 100 V mm^{-1}. The bulk photovoltaic effect, which has an even stronger influence, has been reported to generate fields in excess of 1 kV mm^{-1}, as measured directly in [12] or inferred from holographic measurements in [26, 27]. For larger grating periods, the influence from the diffusion field decreases even more, so that the diffusion field can be ruled out as suitable mechanism for the generation of internal electric fields for POT. For smaller grating lengths, the diffusion field increases significantly, but is then limited by the saturation field in Eq. 3.8, i.e. the number of trapping sites that can contribute to the electric field. It can be easily shown that if $E_{\mathrm{d}} \gg E_{\mathrm{q}}$, the space-charge field can be reduced to $E_{SC} \approx E_q$, which rises linearly at a slope of $E_q/\Lambda = 0.046$V. In this calculation, values of $\varepsilon = 56\varepsilon_0, N_D = 1\times10^{16}\text{cm}^{-3}$ and $N_A = 0.5\times10^{15}\text{cm}^{-3}$ were assumed [13, 28], which are typical for bismuth silicon oxide (BSO). Due to these reasons, the selection of appropriate substrate materials will be reduced to two classes of materials and phenomena: (i) crystals that exhibit a strong bulk photovoltaic effect, for which $LiNbO_3$ will be used as a model system in this thesis [12], or (ii) crystals to which an external field is applied so that drift is the main mechanism for the creation of internal space-charge fields, like it is the case in a BSO crystal [29, 30]. Although the initial contributions to the field of photorefractive opto-electronic tweezers featured the use of lithium niobate to economize the external voltage supply, even the use of a drift-dominated material—which still requires such a voltage supply—can have possible advantages, since the electric field will always be created in-plane and no sandwich structure of electrode/photoconductor/electrode is needed, which leaves the upper surface accessible.

3.4 Bulk Photovoltaic Effect

While the charge transport processes of drift and diffusion can be understood as driven by the physical mechanisms of electrostatic attraction (drift) or a concentration gradient (diffusion), the photovoltaic behavior of some photorefractive materials is more difficult to understand. The term *photovoltaics* is a combination of the greek word for light and the unit for electrical voltage and describes the ability to generate electrical energy upon illumination. This effect is today widely known for its large impact on the production of sustainable electrical energy, mostly by semiconducting photovoltaic cells, but recently also by polymeric compound materials [31]. For the photovoltaic effect in semiconductors, only the charge depletion zone around the contact area of differently doped semiconductors is responsible for the charge separation of holes and electrons [32]. In contrast to this effect, in photorefractive $LiNbO_3$, the bulk of the crystal contributes to the charge transport, which is the reason why the effect is often termed as the *bulk photovoltaic effect* (PVE). Although the main focus in this thesis is on $LiNbO_3$, which shows one of the most pronounced PVE among all photorefractives, it should not go unmentioned that the same effect in weaker form has been observed in a multitude of different materials like $BaTiO_3$, BSO, or $KNbO_3$ [14].

The PVE has been described phenomenologically by Glass et al. in a very early publication, where they found the generated photocurrent j to be proportional to the absorbed optical power [12]

$$j = \kappa\alpha I, \tag{3.12}$$

where α is the absorption coefficient of the lithium niobate crystal and I the incident intensity. κ is a phenomenological constant that describes the proportionality and which in subsequent publications has been renamed to *Glass coefficient* G [14]. Due to the anisotropic nature of the $LiNbO_3$ crystal, the Glass coefficient is not a scalar value, but a third-rank tensor [14]:

$$j_i = \kappa_{ijk}\alpha e_j e_k^* I = \beta_{ijk} e_j e_k^* I, \tag{3.13}$$

where $e_{j,k}$ are the polarization vectors of the incident light waves and β_{ijk} are the components of the photovoltaic tensor that describes the charge transport due to the PVE. The symmetry of the crystallographic *3m* point group, to which $LiNbO_3$ belongs, requires that the photovoltaic tensor has only 4 independent coefficients [33], β_{333}, β_{311}, β_{222} and β_{113}. Like the Pockels effect that will be described in more detail in the upcoming section, the PVE can only occur in non-centrosymmetric systems. Since upon spatial inversion, the photovoltaic current in a centrosymmetric crystal does change its sign, but the product of the incident waves does not, it is required that

$$\beta_{ijk} = -\beta_{ikj}, \tag{3.14}$$

which can only be fulfilled if all elements of the photovoltaic tensor vanish. In $LiNbO_3$, the most prominent charge transport is the linear photovoltaic effect along the crystallographic c-axis of the material, with the coefficients $\kappa_{333} = 2.7 \times 10^{-9}\text{cmV}^{-1}$ and $\kappa_{311} = 3.3 \times 10^{-9}\text{cmV}^{-1}$ being almost equal, whereas the photovoltaic charge transport along the perpendicular axes is significantly reduced, as $\kappa_{222} = 0.3 \times 10^{-9}\text{cmV}^{-1}$ is one order of magnitude lower [33]. Note that for the photovoltaic coefficients, as well as for the electro-optic coefficients in the next section, a contracted notation is used in the literature [13, 30], so that the last two indices are summarized into one.

$$\begin{aligned} (11) &\rightarrow 1 \qquad (23) = (32) \rightarrow 4 \\ (22) &\rightarrow 2 \qquad (31) = (13) \rightarrow 5 \\ (33) &\rightarrow 3 \qquad (12) = (21) \rightarrow 6 \end{aligned}$$

In this new notation, κ_{333} and κ_{311} are reduced to κ_{33} and κ_{31}.

From the previous assumptions of a photoexcitation of charge carriers into the conduction band, a macroscopic charge separation can only be understood in the presence of an external field, which is not the case for $LiNbO_3$ crystals, where open-circuit voltages of several kV mm^{-1} have been measured in iron-doped samples without any externally applied field [26, 27]. It is now a widely accepted microscopic theory that in photovoltaic materials the photoexcited charge carriers receive an additional asymmetric momentum. This momentum is responsible for asymmetric mean photovoltaic drift lengths for movement in the direction of spontaneous polarization or against it [12], which allows the charge carriers to travel in a preferential direction before being recaptured at a different impurity center [34]. In a cumulative effect of all excited charge carriers, this creates the photovoltaic field. The stationary saturation case is reached when the drift of electrons in this field counterbalances the photovoltaic current.

3.5 Visualization of Internal Electric Fields: Pockels Effect

The creation of an internal space-charge field alone is sufficient to obtain electrokinetic forces, but to this end one would not necessarily employ photorefractive materials, but any photoconductive material. In addition to the previously mentioned photoconductivity and the ability to redistribute charge carriers to create electric fields, a photorefractive crystal is electro-optic, which means that its optical parameters are changed by an external or internal electric field. In general, the polarization of any material can be described as a Taylor expansion with respect to the electric field [35]:

$$P_i = \varepsilon_0 \left(\chi^{(1)} E_i + \chi^{(2)}_{ijk} E_j E_k + \chi^{(3)}_{ijkl} E_j E_k E_l + \cdots \right). \tag{3.15}$$

In this expansion, the first term describes the linear change of polarization with the electric field for dielectric materials. However, some materials possess higher order terms, which describe the occurence of the linear $\left(\chi^{(2)}\right)$ and quadratic electro-optic effect $\left(\chi^{(3)}\right)$, also known as *Pockels* and *Kerr effect*, respectively [36]. The modification of the refractive index by linear and quadratic electro-optic effect is typically expressed in terms of changes of the impermeability tensor, which is defined as the inverse of the dielectric tensor [13]

$$\Delta\eta_{ij} = \Delta \left(\frac{1}{n^2} \right)_{ij} = r_{ijk} E_k + s_{ijkl} E_k E_l, \tag{3.16}$$

where r_{ijk} and s_{ijkl} denote the linear and quadratic electro-optic coefficients. Of special interest for this thesis is the linear electro-optic effect. The Kerr effect can be neglected since the intensities of the electric fields that occur are too small. However, for sufficiently high intensities, such as they would be expected in pulsed laser systems, or for media with very high nonlinear coefficient, such as artificial Kerr media, the third order nonlinearity can lead to effects such as self-focusing [37].

Using the Taylor series approximation $\Delta n^{-2} \approx -2n^{-3}\Delta n + \cdots$, and neglecting higher order terms for small changes of the refractive index n, Eq. 3.16 can be written in a more accessible form [38]

$$(\Delta n)_{ij} = -\frac{1}{2} n_{ij}^3 r_{ijk} E_k, \tag{3.17}$$

where n_{ij} describes the unmodified refractive index and r_{ijk} is the electro-optic coefficient. In general, the electro-optic coefficient is a tensor of rank 3 and like the photovoltaic tensor can only have non-vanishing components in crystallographic systems lacking a center of inversion (cf. Eq. 3.14). Note that in the following, the contracted notation as described before will be used. For the two crystal types that will be discussed in this thesis, the only non-vanishing components of the electro-optic tensor can be found in Table 3.1. The results for the coefficients obtained by different

Table 3.1 Electro-optic coefficients for $LiNbO_3$ and BSO in contracted notation at $\lambda = 632.8$ nm

Crystal	Coefficient	Value/pm V^{-1}
$LiNbO_3$	r_{13}	10.0 [39]
		10.9 [40]
–	r_{33}	32.2 [39]
		34.0 [40]
–	r_{22}	6.8 [39]
–	r_{51}	32.6 [39]
BSO	r_{41}	4.4 [41]
		5 [42]

authors depend on the measurement method and ambient conditions and differ by more than 10 %. In this thesis, the values of $r_{13} = 10.0\,\text{pm V}^{-1}$, $r_{33} = 32.2\,\text{pm V}^{-1}$ for $LiNbO_3$ and $r_{41} = 4.4\,\text{pm V}^{-1}$ for BSO were assumed.

For $LiNbO_3$, the very convenient case occurs that two large electro-optic coefficients are along the principal axes of the crystal geometry, namely the polar z/c-axis, along which there is also the most prominent charge transport. Due to the reasons presented above, diffusion fields can be neglected. Without external fields applied to the crystal, it is a safe assumption that $\vec{E} \approx E_z$, so that

$$\Delta n_{o,e} = -\frac{1}{2} n_{o,e}^3 r_{13,33} E_z. \tag{3.18}$$

Here the indices o, e denote *ordinary* (perpendicular to c-axis) and *extraordinary* (parallel to c-axis) polarization.

For the case of BSO, however, the situation is more complicated, because the only non-vanishing electro-optic coefficient is not along one of the principal axes. Therefore, the method of index ellipsoid has to be used to determine the refractive index modification. The index ellipsoid describes the three-dimensional refractive index surface for a given material. Along each direction, the length of a line intersecting this surface determines the refractive index along this direction. The electro-optic effect modifies the respective refractive indices, so that in a general form the ellipsoid can be written as [13]

$$\left(\frac{1}{n_x^2} + r_{1k} E_k\right) x^2 + \left(\frac{1}{n_y^2} + r_{2k} E_k\right) y^2 + \left(\frac{1}{n_z^2} + r_{3k} E_k\right) z^2 + 2yz r_{4k} E_k + 2xz r_{5k} E_k + 2xy r_{6k} E_k = 1. \tag{3.19}$$

In BSO, which belongs to the cubic symmetry class 23, the only non-vanishing coefficients of the electro-optic tensor are $r_{41} = r_{52} = r_{63}$, hence Eq. 3.19 can be simplified to

$$\frac{x^2}{n_x^2} + \frac{y^2}{n_y^2} + \frac{z^2}{n_z^2} + 2xy r_{41} E_z = 1, \tag{3.20}$$

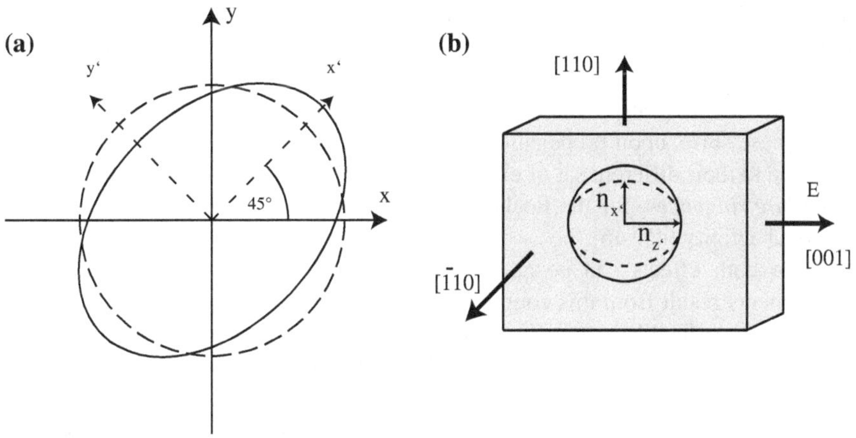

Fig. 3.2 Geometrical parameters for the refractive index modulation in BSO, **a** principle indices of refraction; **b** transverse [$\bar{1}$10] crystal cut

where, without loss of generality, $\vec{E} = E_z = E_3$ was assumed. Turning the reference system by 45° through the coordinate transform of $x = x' \cos 45° - y \sin 45°$, $y = x' \sin 45° + y' \cos 45°$ and $z = z'$, Eq. 3.20 can be rewritten in new principal coordinates and the refractive index modulation is:

$$\Delta n_{x',y'} = \mp \frac{1}{2} n^3 r_{41} E_z. \tag{3.21}$$

The only difference between the two index modulations for $LiNbO_3$ and BSO are that in the case of BSO, the principal eigenindices are along the 45° diagonals between x and y. For the application as holographic devices, it is possible to obtain commercial samples as a [$\bar{1}$10]-cut, where [$\bar{1}$10] indicates the Miller indices that describe the orientation of the crystal plane [43]. In the case of BSO, the samples are cut to be used in the transverse geometry, so that $x' \perp z$ and the refractive index change is described by Eq. 3.21 for light polarized along x' (see [28] and also Fig. 3.2). For light that is polarized parallel to the electric field, the refractive index modulation is zero.

3.5.1 Optical Activity and Pockels Effect in BSO

A factor that makes the treatment of the electro-optic effect even more difficult is the fact that BSO, like all sillenite materials, shows a strong optical activity [42, 44]. Optical activity describes the gyration of the axis of polarization for incident light propagating through the crystal. In contrast to the previous description of the linear electro-optic effect, which was derived in Cartesian coordinates, in the case of an optically active material, these principal coordinate axes are not eigenvectors any more [44]. Due to this reason, right- and left-hand circularly polarized states are introduced as the new eigenvectors, into which a linearly polarized light wave is decomposed. In this way, optical activity alone can be mathematically modeled as *circular birefringence*, where right- and left-handed polarization experience different phase shifts upon propagation, which results in a turning of the polarization axis. Due to their different set of eigenvectors, the two effects are often also termed *linear birefringence*—for the Pockels effect—and *circular birefringence*—instead of optical activity [45, 46].

While both effects can be adequately and easily treated independently, two predicaments result from this coupling of linear and circular birefringence: first of all, even for perfectly linear polarization along one of the principal axes of linear birefringence, the outcoming polarization will always be elliptic. Secondly, the outcoming light wave will in general experience a phase shift that is lower than it would be in a crystal without optical activity [45, 47]. This has to be taken into account when quantitative phase retardations are obtained and discussed in BSO.

3.6 Measurement Techniques for the Evaluation of Photorefractive Media

It could be shown in the previous description that—provided an appropriate geometry is used—the refractive index of both BSO and $LiNbO_3$ crystals is modified by an internal space-charge field upon inhomogeneous illumination. Since this modified property is directly accessible with the optical methods, it can be used for the visualization of an electric field structure as well the measurement of its properties like field magnitude or temporal behavior.

3.6.1 Diffraction Efficiency

A periodic refractive index grating, such as it is described by the harmonic ansatz for the photorefractive grating formation gives rise to diffraction of incident electromagnetic waves. Diffraction phenomena at regular patterns are used still today in grating spectrometers [48] and have yielded structural information about crystals [49], long before this information was directly accessible by scanning tunneling or atomic force microscopy [50, 51]. For photorefractive materials, the diffraction of laser light at the periodic index modulation of the material can be used to obtain information about the magnitude of the electric field. Kogelnik used coupled wave theory to derive a model for the interaction between two beams incident on a photorefractive material (see Fig. 3.3). His derivation of a general solution of coupled modes allows the calculation of the diffracted and transmitted portion of the wave, even for Bragg angle mismatch, slanted gratings or absorption gratings [52].

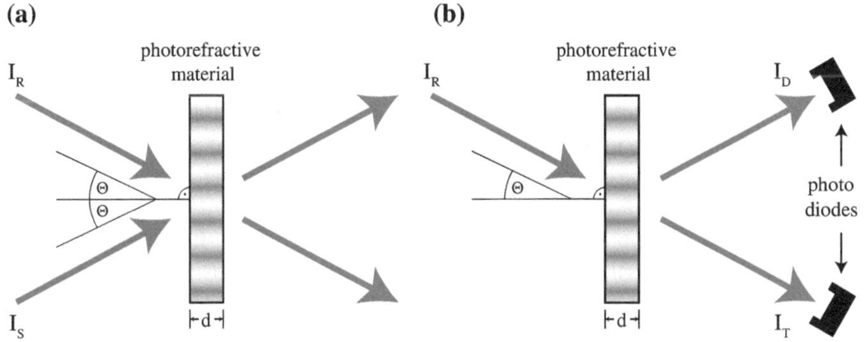

Fig. 3.3 Measurement of the internal field using the diffraction efficiency as derived by Kogelnik: **a** writing of an internal phase grating using the interference of two plane waves; **b** readout of the grating by measuring the transmitted and diffracted portion of a reference beam incident under the Bragg angle

Starting from the scalar wave equation, Kogelnik has solved the coupled-wave equations and obtained an analytical solution for the diffraction efficiency η

$$\eta = \frac{|c_S|}{c_r}\frac{SS^*}{RR^*}, \tag{3.22}$$

where c_r and c_s are obliquity factors introduced to conserve the energy flow and S and R are diffracted wave and reference wave, respectively. The details about the derivation of the ratio between reference and diffracted wave can be found in [52]. The most important result is that for the case of a lossless dielectric grating and perfect Bragg incidence, η becomes the well-known Kogelnik equation [52]

$$\eta = \sin\left(\frac{\pi \Delta n d}{\lambda \cos(\theta)}\right), \tag{3.23}$$

where Δn is the refractive index modulation, d the thickness of the material, θ half the intersection angle between the two incident beams and λ the light's vacuum wavelength. Inverting Eq. 3.23, the refractive index modulation can be calculated from the measurement of the diffraction efficiency and the geometrical parameters of the setup. Rather than normalizing the diffracted wave to the incident reference wave, both the transmitted and diffracted intensity of laser are measured and added up to the reference intensity $\eta = I_d/(I_t + I_d)$ [26]. The measurement of the diffraction efficiency in a holographic setup is the standard technique to determine time constants and magnitude of the space-charge field. The simplest procedure that can be conceived for this technique is to induce a phase grating by interference of two beams and periodically block one of the beams to measure transmitted and diffracted intensity, respectively. More sophisticated setups consist of a writing laser with high power, which is split up into two beams to write a sinusoidal grating in the material and a probe beam of lower power, whose Bragg reflection is measured with two photometers. This approach allows the continuous measurement of the grating strength without erasure since the readout power is negligible in comparison to the strong writing interference pattern. Examples setups can be found in [26, 27]. The latter approach has the advantage that the temporal resolution is not limited by a (mechanical) shutter, but is also more challenging from the experimental point of view, since the second laser must be accurately Bragg-matched to the grating. The quantification of photorefractive media by the technique of holographic diffraction possesses the advantage that it is comparably fast and, apart from very simple arithmetic operations, requires no post-processing of the data. However, since the intensity pattern inducing the grating is formed by interference, special care must be taken of the phase stability of the system. For longer writing times, phase-locking schemes have been developed to correct for the drift of optical components [53]. Even if all conditions have been adequately met, there is no information to be gained about the spatial structure of the internal space-charge field, since the model developed by Kogelnik is based on the assumption of a harmonic modulation and infinitely extended waves [52] and only one scalar value for the diffraction efficiency is measured. Further spatial information can only be acquired by two-dimensional techniques, which will be described in the following.

3.6.2 Zernike Phase Contrast

Since the invention of the microscope in the end of the 16th century, many researchers have used the new tool for investigations in the field of medicine and biology [54]. It has long been a strong drawback that the visibility of objects in biological samples is very poor due to the fact that those samples are often very thin and consist of almost purely phase objects, i.e. objects that absorb only very little light, but induce a small phase retardation to the light transmitted. With respect to these properties, biological samples and photorefractive media are very similar. Unfortunately, the naked eye with its very limited temporal resolution can only detect intensities, which rendered the observation of biological objects hard if not impossible. Dutch physicist Frits Zernike has been motivated by this problem and invented a method to increase the visibility of phase objects in a microscope [55]. His *phase contrast* technique, that he was awarded a Nobel Prize for in 1955, uses an annularly shaped illumination, matched to a phase ring in the back plane of a microscope objective. The phase ring induces a phase shift of $\pi/2$ [56]. All the light directly transmitted through the samples passes through this ring while light that is scattered by the sample bypasses the ring (cf. Fig. 3.4). The scattered light is retarded by $\varphi = 2\pi/\lambda \Delta d \Delta n$ due to the thickness and/or refractive index difference of the sample. For small phase perturbations introduced by the sample, the normalized transmitted wave may be Taylor expanded to [57]

$$e^{i\varphi} \approx 1 + i\varphi(x, y), \tag{3.24}$$

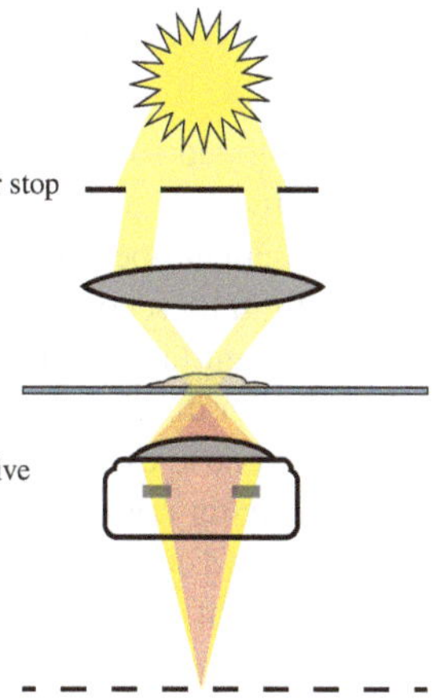

Fig. 3.4 Schematic illustration of Zernike phase contrast

where the first term corresponds to the undiffracted (background) image intensity and the second part contains the two-dimensional phase information. The intensity of the signal would still be equal to one, since the linear phase-modulated terms cancel out and only terms of order φ^2 remain, which in first approximation are negligibly small. If the unmodulated direct light is also phase-shifted by $\pi/2$, both terms can effectively interfere and transfer the phase modulation into an amplitude-modulated image:

$$I(x', y') \approx |\mp i + i\varphi(x, y)| \approx 1 \mp 2\varphi(x, y) \tag{3.25}$$

In general the direct light can be advanced or retarded by $\pi/2$, yielding either *positive* or *negative* phase contrast [57]. Zernike phase contrast shares with common white light illumination the advantage of being very simple and robust. Furthermore, it yields two-dimensional information about the structure of a phase object without any post-processing. However, in the above derivation, some strong approximations have been made, and in general the relation between phase and intensity value is only linear for a small values of the phase retardation [58]. Of course, higher phase retardations could be measured using a look-up-table, but this calibration would have to be done beforehand and is highly dependent on the illumination conditions of the microscope. Furthermore, the acquisition of quantitative data is further impeded by the occurrence of artifacts like halo or shade-off [56]. Therefore, whenever a space-charge field is induced in a photorefractive crystal, Zernike phase contrast is used in this thesis for quickly checking the position and quality of the refractive index modulation, but not for the quantification of the amplitude of the internal electric field.

3.6.3 Digital Holographic Microscopy (DHM)

A third method that can be employed for two-dimensional mapping of the internal electric fields inside a photorefractive material is *digital holographic microscopy* (DHM) [59–61], where mathematical operations are used to recover both amplitude and phase information from a digitally recorded interferogram. Unlike Zernike phase contrast, this techniques gives quantitative values for the phase retardations of light without any calibration. One earlier disadvantage of this technique was that some of the operations necessary for the phase extraction were computationally quite intensive. However, nowadays with rising computer power and more advanced libraries to efficiently perform many standard operations, the numerical calculation of a hologram can be achieved in fractions of a second. More details about the performance of the DHM acquisition will be given in Sect. 3.6.3.3.

The basis of every DHM acquisition is an interferogram between a signal wave, which carries the information about the phase object, and a constant reference wave. To arrive at such an interferogram, different methods exist, among which the most straightforward is to split up a laser beam in signal and reference wave with two spatially separated beam paths in a Mach-Zehnder interferometer [62–64]. In order to separate twin-image and 0th order undiffracted light, both waves have to interfere at an angle, leading to the occurrence of interference fringes on the recording camera, whose curvature contains the phase information. Due to the fact that one of the waves propagates at an angle to the optical axis, this method is often referenced to as *off-axis holography* [65]. Another approach to digital holography, which makes use of the interference between scattered and unscattered light, is—in contrast to the previously mentioned setups—often termed *in-line* or *on-axis holography*, since both reference and signal wave propagate on the same optical axis [66, 67]. Although the optical setup is in principle much simpler, a strong disadvantage is the occurrence of the holographic twin-image, for whose suppression additional algorithms have to be applied [68]. For the reasons presented above and the efficient algorithms for the calculation of phase values, in this thesis the off-axis holographic setup is used in a Mach-Zehnder configuration.

3.6.3.1 DHM Setup for the Assessment of Photorefractive Crystals

Figure 3.5 depicts the basic setup for the measurement of phase retardations in photorefractive crystals. With minor modifications, it will be used wherever quantitative phase data is obtained in this thesis. In general, it can be divided into a low-power readout part and a high-power writing part that contains a spatial light modulator.

The first part of the setup consist of a Mach-Zehnder interferometer, where a helium-neon laser (HeNe, $\lambda = 632.8$ nm) is spatially filtered, expanded and split up at a polarizing beam splitter (PBS) into signal and reference wave. The intensity ratio between both beams can be tuned by a half-wave plate (HWP) before the beam splitter. The signal beam is used to illuminate the photorefractive sample (S) under investigation. The image of the crystal is magnified by an infinity-corrected microscope objective (MO) and combined with the reference wave at a 50/50 beam splitter (BS). A half-wave plate in the reference arm turns the polarization of the reference beam, so that light from both signal and reference arm can interfere. A tube lens (TL) generates the image on the CCD/CMOS camera. A long-pass color filter (CF) just before the camera suppresses any remaining green light from the light-induced structuring of the material. In some variations of the above setup, a second microscope objective is introduced also in the reference path. Although such an objective is not absolutely necessary for the measurement, it is inserted for convenience to establish a similar optical path for both beams. Otherwise a different wave front curvature of both beams will introduce a circular phase offset that has to be corrected for [69].

The second part of the setup is the *writing arm* for the optical induction of internal electric fields. Whenever in the following electric fields are "written" or "induced", this is a reference to this green beam path. As described in the chapter about the photorefractive processes, a modulated internal space-charge field inside a photorefractive crystal will form only in the presence of an inhomogeneous illumination. The easiest conceivable way to induce such a modulation is via the interference of two plane waves. However, to be more flexible and to have full electronic pattern control, an amplitude spatial light modulator (ASLM, HED 6001, *HoloEye*, Berlin) is employed for the induction of arbitrary intensity patterns. It consists of a Full-HD vertically aligned nematic display, whose pixels can be independently addressed with 8-bit resolution to turn the input polarization of the incoming light. Due to the nonlinear birefringence properties of the vertically aligned nematic liquid crystals [70], the amplitude modulation transfer function (cf. Fig. 3.5) is nonlinear, but can be easily linearized based on a calibrated look-up table. Using this technique, arbitrary binary patterns can be induced in the photorefractive material without any pre-processing. For the generation of patterns with continuous intensity gradients, the desired pattern has to be matched to the modulator gray values using the inverse of the amplitude transfer function. A frequency-doubled Nd:YAG laser ($\lambda = 532$ nm) is spatially filtered and expanded to illuminate the ASLM. A polarizing beam splitter cube or a polaroid filter (P) converts the polarization matrix into an amplitude matrix. In this thesis, a polarizing beam splitter cube is used due its low absorption. The image of the ASLM is demagnified by two lenses and introduced into the signal arm of the MZ-interferometer by a dichroic mirror (DM). The green light that is transmitted through the sample is blocked by the previously mentioned color filter. In order to minimize the diffraction of the green light at the refractive index pattern it induces, the writing polarization is always ordinary, while the readout polarization can be adjusted to be either ordinary, which results in smaller phase shifts and less phase-jumps, or extraordinary, for higher phase shifts and better sensitivity.

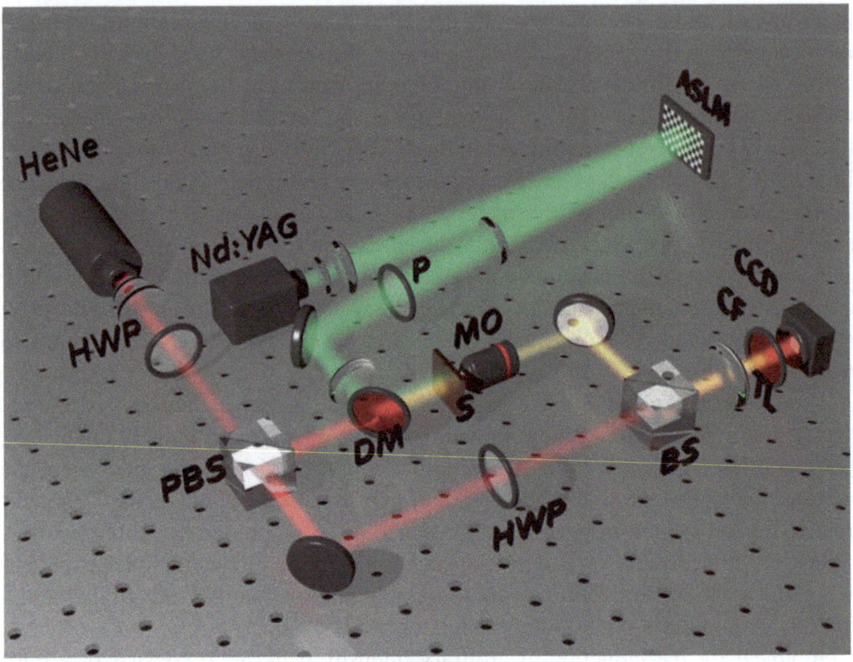

Fig. 3.5 Induction and digital holographic measurement of arbitrary patterns on photorefractive crystals; *ASLM* amplitude spatial light modulator, *CCD* CCD camera, *CF* color filter, *DM* dichroic mirror, *HeNe* helium-neon laser, *HWP* half-wave plate, *MO* microscope objective, *Nd:YAG* Nd:YAG laser, *P* polarizer, *(P)BS* (polarizing) beam splitter, *S* sample, *TL* tube lens; for details about the setup refer to Sect. 3.6.3.1

3.6.3.2 Off-axis Digital Holographic Reconstruction

The reconstruction of phase values in off-axis geometry starts with the acquisition of an interferogram whose intensity can be described as [62]

$$\begin{aligned} I(x,y) &= |A_\mathrm{R}|^2 + |A_\mathrm{S}(x,y)|^2 + 2A_\mathrm{R}A_\mathrm{S}(x,y)\cos\left(q_x x + q_y y + \varphi(x,y)\right) \\ &= I_\mathrm{R} + I_\mathrm{S} + u(x,y), \end{aligned} \tag{3.26}$$

where $A_{\mathrm{R,S}}$ denotes the amplitude of signal and reference wave, respectively, $\vec{q} = \vec{e}_x q_x + \vec{e}_y q_y$ is the grating vector of the interference fringes and $\varphi(x,y)$ is the phase retardation due to the internal space-charge field E_SC that is to be determined. In the case of a weakly absorbing phase object, the signal amplitude can be approximated as constant. Although some of the crystals under investigation, especially $LiNbO_3$ samples, possess quite a significant absorption, the above approximation still holds since this absorption is independent of the refractive index modulation and thus homogeneous over the whole crystal surface. The ratio between signal and reference wave can be tuned by the first half-wave plate in the DHM setup to give maximum interference on the CCD camera.

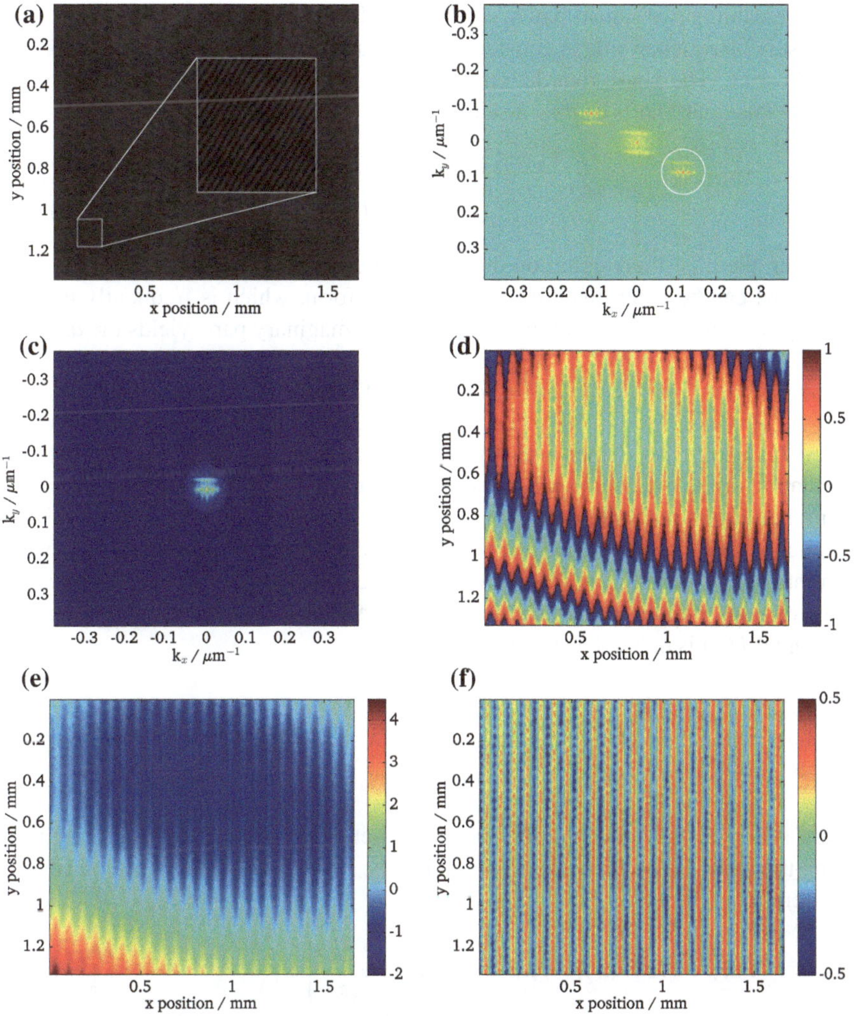

Fig. 3.6 Off-axis digital holographic phase calculation: The interferogram (**a**) is Fourier transformed; in the resulting Fourier spectrum (**b**), only the positive frequency components are filtered as indicated by the *white circle* and shifted back to the center of the spectrum (**c**) to correct for the linear phase gradient; the wrapped phase (**d**) is unwrapped (**e**) and corrected for the background curvature of the interference pattern to reveal the true phase modulation (**f**). Note that all *colorbars* are scaled in units of π

The interferogram is Fourier transformed, where the third term in Eq. 3.26 leads to a convolution of the sinusoidally modulated carrier fringes with the image information and thus to two additional side maxima clearly separated from the constant zero order term provided that the grating vector $\vec{q}$ is large enough (see also Fig. 3.6). The zero order term in the spectrum, which contains only the first two (constant) intensity terms in Eq. 3.26 can be suppressed by simple Fourier filtering.

Of the remaining two side maxima, which result from the third term in Eq. 3.26, only the positive frequency side is considered, the reason for which will be discussed in the following. The phase retardation of the signal wave can be calculated from the real and imaginary parts of the carrier signal by

$$\varphi(x, y) = \arctan\left(\frac{\Im(u(x, y))}{\Re(u(x, y))}\right). \tag{3.27}$$

However, the measured signal on the CCD is purely real. The imaginary part of a real signal can be determined by a Hilbert transform, which is frequently used in signal processing [71]. Combining both real and imaginary parts yields the *analytic signal*:

$$\widehat{I}(x) = I(x) + i\mathcal{H}\{I(x)\}, \tag{3.28}$$

where $\mathcal{H}$ denotes the Hilbert transform. It will be shown in the following that the calculation of the imaginary part does not rely on a computationally intensive algorithm, but occurs naturally during the process of image processing. For the sake of simplicity, the following derivation is limited to the one-dimensional case. Its extension into two dimensions is straightforward, since there is no coupling between the two directions x and y in Cartesian coordinates. By definition, the Hilbert transform of a signal $I(x)$ is defined as [72]:

$$\mathcal{H}\{I(x)\} = \frac{1}{\pi}\int_{-\infty}^{\infty} \frac{I(x)}{x' - x} dx = I(x') * \frac{1}{\pi x'}, \tag{3.29}$$

where $*$ denotes the convolution operation. Let $I(x) \circ\!\!-\!\!\bullet F(\omega)$ be a Fourier pair. Using the convolution theorem, Eq. 3.28 can be rewritten as a sequence of Fourier transforms:

$$\begin{aligned} I(x) + i\mathcal{H}\{I(x)\} &= I(x) + i\left\{I(x) * \frac{1}{\pi x}\right\} \\ &= I(x) + i\mathcal{F}^{-1}\left\{\mathcal{F}\{I(x)\}\mathcal{F}\left\{\frac{1}{\pi x}\right\}\right\} \\ &= \mathcal{F}^{-1}\{F(\omega)\} + i\mathcal{F}^{-1}\{F(\omega)(-i\ sgn(\omega))\} \\ &= \mathcal{F}^{-1}\{F(\omega)\} + \mathcal{F}^{-1}\{F(\omega)sgn(\omega)\} \\ &= \mathcal{F}^{-1}\{F(\omega) + F(\omega)sgn(\omega)\} \end{aligned} \tag{3.30}$$

To arrive at this result, the Fourier transform of the Hilbert operator

$$\frac{1}{\pi y} \circ\!\!-\!\!\bullet - i\ sgn(\omega) \tag{3.31}$$

has been used [72], where $sgn(\omega)$ is the signum function which is 1 for $\omega > 0$ and -1 for $\omega < 0$. It is obvious that the term

$$F(\omega) + F(\omega)sgn(\omega) \begin{cases} 0 & \text{for } \omega < 0 \\ F(\omega) & \text{for } \omega = 0 \\ 2F(\omega) & \text{for } \omega > 0 \end{cases} \tag{3.32}$$

describes nothing else than the previously introduced Fourier operations, where the DC term is set to zero anyway and only the positive frequency spectrum is kept for further calculations. The fact that the result of the inverse Fourier transform is scaled by a factor of 2 can be neglected for the calculation of the phase of a signal, since the additional factors cancel out anyway in Eq. 3.27. Should the absolute value of the real and/or imaginary part be of interest however, they can be obtained by multiplying the result of $\mathcal{F}^{-1}\{2F(\omega)\}$ by a factor of $1/2$ due to the linear properties of the Fourier transform. The two-dimensional phase shift $\varphi(x, y)$ (see Eq. 3.27) can then be calculated for each pixel and allows a determination of the space-charge field via Eq. 3.17

$$E_{\mathrm{SC}}(x, y) = -\frac{\Delta\varphi(x, y)\lambda}{\pi d n^3 r_{\mathrm{eff}}}, \tag{3.33}$$

where d is the thickness of the photorefractive crystal, r_{eff} the effective electro-optic coefficient in the respective geometry and $\Delta\varphi(x, y) = \varphi(x, y) - \varphi_0$ is the relative phase shift to an arbitrary offset reference phase φ_0. The total workflow is depicted in more detail in Fig. 3.6.

3.6.3.3 Origin of Phase Jumps and Two-Dimensional Phase Unwrapping

It has been shown that the calculation of the imaginary part of the signal beam occurs naturally during the process of Fourier filtering. Since the phase φ is obtained using the two-dimensional `arctan2()` function, whose results by definition are limited to the interval $[-\pi, \pi)$, the resulting phase may be strongly wrapped, which means that phase jumps of 2π occur whenever the phase retardation leaves the defined range of values of $-\pi \leq \varphi < \pi$. To arrive at meaningful result even for phase shifts that exceed 2π, the phase information has to be unwrapped, i.e. the phase jumps have to identified and corrected for by adding multiples of 2π to arrive at a continuous phase map. The problem of *phase unwrapping* in the two-dimensional case can be a very challenging task that finds applications in various fields of research, like the interpretation of radar data as well as in magnetic resonance imaging [73, 74]. There are several algorithms that can be applied to unwrap the two-dimensional phase information and they all differ in quality as well as well as computational time [73]. The algorithms that were implemented and tested in the course of this thesis can be found in Appendix A, where also a discussion will be presented about the origin of unwrapping errors.

The phase jumps in the measurements originate from two processes; the phase retardation of the object and a linear phase gradient from the term $\vec{q} \cdot \vec{r}$. This linear gradient is a mere artifact resulting from the off-axis geometry and must be removed to reveal the interesting phase retardation. To remove the linear gradient, spatial frequencies that are cut out around the first side maximum are shifted to the DC region of the spectrum, i.e. the center of the image in the Fourier plane before doing the inverse Fourier transform (cf. Fig. 3.6), which is equivalent to removing the gradient in the image domain [69]. Once the linear gradient is removed, the quantitative phase retardation under investigation is revealed. By this procedure, many phase jumps from the strong linear gradient have been avoided in the first place, however, it should not go unmentioned that only if the two arms of the Mach-Zehnder interferometer are perfectly aligned and both wavefronts have the exact same curvature, the interference fringes are perfectly straight and the resulting phase profile after linear gradient removal is a flat profile. Due to small misalignments and abberations in the path of the beam, in most of the cases the resulting phase profile will be overlayed with a curved offset phase that can lead to additional phase jumps. Even if such abberations can be corrected for by assuming a quadratic phase curvature [69], in the case of a time-resolved measurement of the phase retardation inside a photorefractive crystal there is always the possibility to define a reference background phase where no internal electric fields are present. After the calculation and unwrapping of all phase images, this offset curvature can be subtracted to reveal only the phase modulations due to the internal electric field (cf Fig. 3.6f).

Last but not least, for sufficiently high space-charge fields, the phase retardation due to the electro-optic effect may still be larger than 2π. For BSO samples, this is not so much of a problem, since the externally applied fields in air are technically limited by the ionization voltage of air, which depends strongly on the level of humidity, but lies in the range of about 3 kV mm^{-1} [75, 76]. Equation 3.33 allows to estimate that the phase shift occuring in BSO in this scenario would be approximately 0.39 πmm^{-1}, so even for crystals of a few mm, there would be most likely no phase jump due to wrapped phases. On the other hand, in the case of lithium niobate, with the electro-optic coefficient being larger and with internal fields that have already been measured to be as high as and even exceed 10 kV mm^{-1}, the phase shift can be up to $\varphi \approx 1.80\ \pi\text{mm}^{-1}$ for ordinary polarization and even 5.05 πmm^{-1} for extraordinary polarization, so a phase jump is very likely to occur for sufficiently large crystals.

In the discussion of different techniques to measure the magnitude of internal fields, it has been mentioned that the Kogelnik method of determining the diffraction efficiency and as well as Zernike phase contrast need no or negligible postprocessing and are hence computationally very fast. In contrast to these methods, DHM yields quantitative as well as structural information at the expense of computational time. In the course of this thesis, all image acquisition and processing, as well as the DHM calculation, were performed with MATHWORKS MATLAB. The process of calculating the phase of an interferogram involves two Fourier transform operations of, in this case, 1280×1024 pixel images, the manipulation of the Fourier spectrum, the calculation of the `arctan2()` for each pixel and, if necessary, the unwrapping procedures. Due to optimized matrix operations for Fast Fourier transform (FFT) [77], the two-dimensional FFT for an image of 1280 × 1024 pixels takes approximately 100 ms for each FFT, and 200 additional milliseconds for all additional Fourier manipulations and matrix operations except the two-dimensional unwrapping (checked on an Intel Core 2 Duo, running at 3 GHz core speed). The main part of each calculation is consumed by the unwrapping procedure, which in the best case can be accomplished in a another 400 ms. Details about the problems in two-dimensional phase unwrapping and the algorithms that have been developed in this thesis can be found in Appendix A. Although the calculation of the internal electric space charge field takes approximately 1 s for each timestep, this drawback is by far compensated by the additional information that is to be gained about the structure of the internal fields, as will be demonstrated in the following chapters.

References

1. C.F. Gonzalez, V.T. Remcho, Fabrication and evaluation of a ratchet type dielectrophoretic device for particle analysis. J. Chromat. A **1216**(52), 9063–9070 (2009)
2. P. Chiou, A. Ohta, M. Wu, Massively parallel manipulation of single cells and microparticles using optical images. Nature **436**, 370–372 (2005)
3. M.C. Wu, Optoelectronic tweezers. Nat. Photonics **5**(6), 322–324 (2011)
4. M. Esseling, F. Holtmann, M. Woerdemann, C. Denz, Two-dimensional dielectrophoretic particle trapping in a hybrid crystal/PDMS-system. Opt. Express **18**(16), 17404–17411 (2010)
5. S. Glaesener, M. Esseling, C. Denz, Multiplexing and switching of virtual electrodes in optoelectronic tweezers based on lithium niobate. Opt. Lett. **37**(18), 3744–3746 (2012)
6. A. Ashkin, G. Boyd, J. Dziedzic, R. Smith et al., Optically-induced refractive index inhomogeneities in $LiNbO_3$ and $LiTaO_3$. Appl. Phys. Lett. **9**, 72–74 (1966)
7. F. Chen, Optically induced change of refractive indices in $LiNbO_3$ and $LiTaO_3$. J. Appl. Phys. **40**(8), 3389 (1969)
8. R. Townsend, J. LaMacchia, Optically induced refractive index changes in batio3. J. Appl. Phys. **41**, 5188–5192 (1970)
9. G. Lesaux, G. Roosen, A. Brun, Observation and analysis of the fast photorefractive process in BSO. Opt. Commun. **56**(6), 374–378 (1986)
10. L. Yu, W. Chan, Z. Peng, A. Gharavi, Multifunctional polymers exhibiting photorefractive effects. Acc. Chem. Res. **29**(1), 13–21 (1996)
11. K. Buse, Light-induced charge transport processes in photorefractive crystals II: materials. Appl. Phys. B Lasers Opt. **64**(4), 391–407 (1997)
12. A. Glass, D. Linde, T. Negran, High-voltage bulk photovoltaic effect and the photorefractive process in $LiNbO_3$. Appl. Phys. Lett. **25**(4), 233–235 (1974)
13. P. Yeh, *Introduction to Photorefractive Nonlinear Optics* (Wiley, New York, 1993)
14. B. Sturman, V. Fridkin, *The Photovoltaic and Photorefractive Effects in Noncentrosymmetric Materials* (Gordon and Breach Science, New York, 1992)
15. N. Kukhtarev, V. Markov, S. Odulov, M. Soskin et al., Holographic storage in electrooptic crystals 1: steady state. Ferroelectrics **22**, 949–960 (1979)
16. N. Kukhtarev, V. Markov, S. Odulov, M. Soskin et al., Holographic storage in electrooptic crystals 2: beam coupling—light amplification. Ferroelectrics **22**, 961–964 (1979)
17. M. Clark, F. Disalvo, A. Glass, G. Peterson, Electronic structure and optical index damage of iron-doped lithium-niobate. J. Chem. Phys. **59**(12), 6209–6219 (1973)
18. D. Bryan, R. Rice, R. Gerson, H. Tomaschke et al., Magnesium-doped lithium-niobate for higher optical power applications. Opt. Eng. **24**(1), 138–143 (1985)

19. O. Schirmer, O. Thiemann, M. Wohlecke, Defects in $LiNbO_3$-I. Experimental aspects. J. Phys. Chem. Solids **52**(1), 185–200 (1991)
20. B. Wechsler, M. Klein, C. Nelson, R. Schwartz, Spectroscopic and photorefractive properties of infrared-sensitive Rhodium-doped barium-titanate. Opt. Lett. **19**(8), 536–538 (1994)
21. K. Peithmann, J. Hukriede, K. Buse, E. Kratzig, Photorefractive properties of LiNbO3 crystals doped by copper diffusion. Phys. Rev. B: Condens. Matter Mater. Phys. **61**(7), 4615–4620 (2000)
22. A. Egorysheva, V. Burkov, Y. Kargin, A. Vasil'ev et al., Spectroscopic properties of Bi12SiO20 and Bi12OTiO20 crystals doped with Mn, Mn+P, Cr+P, Cr+Ga, and Cr+Cu. Inorg. Mater. **37**(8), 817–824 (2001)
23. M. Carrascosa, F. Agullo-Lopez, Erasure of holographic gratings in photorefractive materials with 2 active species. Appl. Opt. **27**(14), 2851–2857 (1988)
24. M. Bashaw, T. Ma, R. Barker, S. Mroczkowski et al., Theory of complementary holograms arising from electron-hole transport in photorefractive media. J. Opt. Soc. Am. B: Opt. Phys. **7**(12), 2329–2338 (1990)
25. M. Bashaw, M. Jeganathan, L. Hesselink, Theory of 2-center transport in photorefractive media for low-intensity continuous-wave illumination in the quasy-steady-state-limit. J. Opt. Soc. Am. B: Opt. Phys. **11**(9), 1743–1757 (1994)
26. K. Peithmann, A. Wiebrock, K. Buse, Photorefractive properties of highly-doped lithium niobate crystals in the visible and near-infrared. Appl. Phys. B: Lasers Opt. **68**(5), 777–784 (1999)
27. A. Zaltron, M. Bazzan, N. Argiolas, M.V. Ciampolillo et al., Depth-resolved photorefractive characterization of lithium niobate doped with iron by thermal diffusion. Appl. Phys. B: Lasers Opt. **108**(3), 657–663 (2012)
28. M. Petrov, V. Bryksin, Space-charge waves in sillenites: rectification and second-harmonic generation, in *Photorefractive Materials and Their Applications 2: Materials*, ed. by P. Günter, J.-P. Huignard (Springer, Berlin, 2007), pp. 285–325
29. S. Stepanov, M. Petrov, Efficient unstationary holographic recording in photorefractive crystals under an external alternating electric-field. Opt. Commun. **53**(5), 292–295 (1985)
30. P. Günter, J. Huignard, *Photorefractive Materials and Their Applications 2* (Springer, Berlin, 2007)
31. G. Yu, J. Gao, J. Hummelen, F. Wudl et al., Polymer photovoltaic cells—enhanced efficiencies via a network of internal donor-acceptor heterojunctions. Science **270**(5243), 1789–1791 (1995)
32. D. Chapin, C. Fuller, G. Pearson, A new silicon p–n junction photocell for converting solar radiation into electrical power. J. Appl. Phys. **25**(5), 676–677 (1954)
33. H. Festl, P. Hertel, E. Kratzig, R. von Baltz, Investigations of the photo-voltaic tensor in doped $LiNbO_3$. Phys. Status Solidi A **113**(1), 157–164 (1982)
34. H. Heyszenau, Electron-transport in bulk photo-voltaic effect. Phys. Rev. B: Condens. Matter Mater. Phys. **18**(4), 1586–1592 (1978)
35. Y. Shen, *The Principles of Nonlinear Optics* (Wiley, New York, 1984)
36. B. Saleh, M. Teich, *Fundamental of Photonics* (Wiley Interscience, New York, 2007)
37. A. Ashkin, J. Dziedzic, P. Smith, Continuous-wave self-focusing and self-trapping of light in artificial Kerr media. Opt. Lett. **7**(6), 276–278 (1982)
38. J. Nye, *Physical Properties of Crystals* (Oxford Science, Oxford, UK, 1985)
39. J. Zook, D. Chen, G. Otto, Temperature dependence and model of electro-optic effect in $LiNbO_3$. Appl. Phys. Lett. **11**(5), 159–161 (1967)
40. K. Onuki, N. Uchida, T. Saku, Interferometric method for measuring electrooptic coefficients in crystals. J. Opt. Soc. Am. **62**(9), 1030–1032 (1972)
41. A. Grunnet-Jepsen, I. Aubrecht, L. Solymar, Investigation of the internal field in photorefractive materials and measurement of the effective electrooptic coefficient. J. Opt. Soc. Am. B: Opt. Phys. **12**(5), 921–929 (1995)
42. R. Aldrich, S. Hou, M. Harvill, Electrical and optical properties of Bi12SiO20. J. Appl. Phys. **42**, 493–494 (1971)
43. C. Kittel, *Introduction to Solid State Physics* (Wiley, New York, 1963)

44. A. Yariv, P. Yeh, *Optical Waves in Crystals* (Wiley, New York, 1984)
45. P. Pellatfinnet, Measurement of the electro-optic coefficient of BSO crystals. Opt. Commun. **50**(5), 275–280 (1984)
46. M. Henry, S. Mallick, D. Rouede, L. Celaya et al., Propagation of light in an optically-active electrooptic crystal of BI12SIO20—measurement of the electrooptic coefficient. J. Appl. Phys. **59**(8), 2650–2654 (1986)
47. P. Foote, T. Hall, Influence of optical-activity on 2 beam coupling-constants in photorefractive BI12SIO20. Opt. Commun. **57**(3), 201–206 (1986)
48. J. den Herder, A. Brinkman, S. Kahn, G. Branduardi-Raymont et al., The reflection grating spectrometer on board XMM-Newton. Astron. Astrophys. **365**(1), L7–L17 (2001)
49. W. Friedrich, P. Knipping, M. Laue, Interferenzerscheinungen bei Röntgenstrahlen. Ann. Phys. **4**, 971–988 (1913)
50. G. Binnig, H. Rohrer, C. Gerber, E. Weibel, Tunneling through a controllable vacuum gap. Appl. Phys. Lett. **40**(2), 178–180 (1982)
51. G. Binnig, C. Quate, C. Gerber, Atomic force microscopy. Phys. Rev. Lett. **56**(9), 930–933 (1986)
52. H. Kogelnik, Coupled wave theory for thick hologram gratings. Bell Syst. Tech. J. **48**, 2909–2947 (1969)
53. S. Breer, K. Buse, K. Peithmann, H. Vogt et al., Stabilized recording and thermal fixing of holograms in photorefractive lithium niobate crystals. Rev. Sci. Instrum. **69**(4), 1591–1594 (1998)
54. E. Hecht, *Optik* (Oldenbourg, Munich, 2002)
55. F. Zernike, How I discovered phase contrast. Science **121**, 345–349 (1955)
56. D. Murphy, M. Davidson, *Fundamentals of Light Microscopy and Electronic Imaging*, 2nd edn. (Wiley-Blackwell, New York, 2013)
57. J.W. Goodman, *Introduction to Fourier Optics*, 3rd edn. (Roberts and Company, Greenwood Village, Colorado, 2005)
58. J. Glückstad, D. Palima, *Generalized Phase Contrast* (Springer, Dordrecht, 2009)
59. U. Schnars, W. Juptner, Direct recording of holograms by a CCD target and numerical reconstruction. Appl. Opt. **33**(2), 179–181 (1994)
60. E. Cuche, F. Bevilacqua, C. Depeursinge, Digital holography for quantitative phase-contrast imaging. Opt. Lett. **24**(5), 291–293 (1999)
61. M. de Angelis, S. De Nicola, A. Finizio, G. Pierattini et al., Two-dimensional mapping of electro-optic phase retardation in lithium niobate crystals by digital holography. Opt. Lett. **30**, 1671–1673 (2005)
62. T. Ikeda, G. Popescu, R. Dasari, M. Feld, Hilbert phase microscopy for investigating fast dynamics in transparent systems. Opt. Lett. **30**(10), 1165–1167 (2005)
63. M. Esseling, S. Glaesener, F. Volonteri, C. Denz, Opto-electric particle manipulation on a bismuth silicon oxide crystal. Appl. Phys. Lett. **100**(16), 161903 (2012)
64. M. Esseling, A. Zaltron, N. Argiolas, G. Nava et al., Highly reduced iron-doped lithium niobate for opto-electronic tweezers. Appl. Phys. B: Lasers Opt. **113**(2), 191–197 (2013)
65. E. Cuche, P. Marquet, C. Depeursinge, Spatial filtering for zero-order and twin-image elimination in digital off-axis holography. Appl. Opt. **39**(23), 4070–4075 (2000)
66. W. Xu, M. Jericho, I. Meinertzhagen, H. Kreuzer, Digital in-line holography of microspheres. Appl. Opt. **41**(25), 5367–5375 (2002)
67. B. Javidi, D. Kim, Three-dimensional-object recognition by use of single-exposure on-axis digital holography. Opt. Lett. **30**(3), 236–238 (2005)
68. K. Nugent, Twin-image elimination in Gabor holography. Opt. Commun. **78**(3–4), 293–299 (1990)
69. B. Kemper, G. von Bally, Digital holographic microscopy for live cell applications and technical inspection. Appl. Opt. **47**(4), A52–A61 (2008)
70. Holoeye AG, HED 6001 Display Webpage (2013)
71. U. Kiencke, H. Jäkel, *Signale und Systeme* (Oldenbourg, Munich, 2002)

72. J. Ohm, H. Lüke, *Signalübertragung—Grundlagen der Digitalen und Analogen Nachrichtenübertragungssysteme* (Springer, Berlin, 2007)
73. D.C. Ghiglia, M.D. Pritt, *Two-Dimensional Phase Unwrapping: Theory, Algorithms and Software* (Wiley, New York, 1998)
74. R. Cusack, N. Papadakis, New robust 3-D phase unwrapping algorithms: application to magnetic field mapping and undistorting echoplanar images. Neuroimage **16**(3, 1), 754–764 (2002)
75. D. Meschede, *Gerthsen Physik* (Springer, Berlin, 2001)
76. P. Tipler, G. Mosca, *Physik* (Elsevier, Munich, 2007)
77. J. Cooley, J. Tukey, An algorithm for machine calculation of complex Fourier series. Math. Comput. **19**(90), 297 (1965)

Chapter 4
Quantitative Investigation of Photorefractive Substrate Materials

The upcoming chapter will be concerned with the appropriate selection of substrate materials for the application as optoelectronic tweezers. Very much as it is well-known in daily life, *nothing comes for free*, so there is always a trade-off between certain aspects. This section will discuss the basic properties of two different classes of materials. As pointed out in Sect. 3.3, only materials with a substantial bulk photovoltaic behavior or drift-dominated materials can be considered as good materials for their use as POT. For both material classes, one example substance will be thoroughly investigated, $LiNbO_3$ as a prominent example of the internal photovoltaic effect, and BSO as a representative for materials where drift is exploited for charge transport. It will be discussed how the two material classes under investigation can be modified to obtain best results for DEP. There are two figures-of-merit on the route towards the optimal substrate material, a high electric field and a sufficiently fast response of the material. It will be shown that this optimization process has to be approached from different sides for each material: while $LiNbO_3$ scores with high internal fields, it lacks a quick temporal response. On the other side, the internal fields in BSO are generated quickly due to a high photoconductivity, but the internal field strength is inherently limited to the magnitude of the externally applied field, which is rather low compared to $LiNbO_3$, so a method has to be found to enhance this property.

4.1 Highly Reduced Iron-Doped Lithium Niobate ($LiNbO_3$)

4.1.1 General Properties of $LiNbO_3$

Among the materials that have been used for photorefractive applications, $LiNbO_3$ takes an outstanding position. This material, which is not found in nature, was first grown in the laboratory in 1928 by Zachariasen, and its remarkable ferroelectric properties were first described in 1949 by Matthias and Remeike [1]. It was also in this material that the photorefractive effect was first discovered [2]. The

M. Esseling, *Photorefractive Optoelectronic Tweezers and Their Applications*,
Springer Theses, DOI 10.1007/978-3-319-09318-5_4

very high internal fields in $LiNbO_3$ are generated by the previously described bulk photovoltaic effect (cf. Sect. 3.4), which is responsible for the uni-directional separation of charges even without the presence of an external field. In contrast to several other materials that also show the same effect [3], $LiNbO_3$ has the remarkable property of being able to generate internal field strengths of up to several kV mm^{-1} [4]. Together with its relatively high electro-optic coefficients of $r_{13/33} = 10.0/32.2$ pmV^{-1} and its low photoconductivity, $LiNbO_3$ has been the material of choice for applications in the field of long-term holographic data storage [5, 6]. For the application as POT, as mentioned earlier, the most important figure-of-merit is the highly modulated internal electric field, capable of polarizing and trapping matter on the surface.

During the course of this thesis, several $LiNbO_3$ crystal samples of different size and composition are used. Table 4.1 gives a complete overview over the sample names, their dimensions and reduction degrees, i.e. the ratio between filled and empty, ionized electron traps in Kukhtarev's band transport model (cf. Sect. 3.2). To enhance the photorefractive response in the visible, $LiNbO_3$ is doped with iron impurities during the growth process. Since this material exists only in the two valence states of Fe^{2+} and Fe^{3+}, acting as donor and acceptor centers, the reduction degree of the material is defined as the ratio between their concentrations $c_{Fe^{2+}}$ and $c_{Fe^{3+}}$. More information about the reduction degree will be given later. The total iron concentration is $c_{Fe} = 18.8 \times 10^{18}$cm^{-3} for all samples, since this value was proven to be the optimal concentration. For lower concentrations of trapping centers, no substantial internal field E_{SC} can be formed. On the other hand, for iron dopant levels above the optimal concentration, the refractive index modulation, hence the internal field, was shown to saturate or even decrease, caused by an increased dark conductivity [4].

Table 4.1 List of $LiNbO_3$ crystals used in this thesis, their dimensions and reduction degrees; the total iron concentration for all crystals is $c_{Fe} = 18.8 \times 10^{18}$cm^{-3}

	Reduction degree r	Dimensions (mm)
AT1	0.039 ± 0.004 (as-grown)	$10 \times 10 \times 0.50$
AT2	0.039 ± 0.004	$15 \times 13 \times 2$
151.1	0.08 ± 0.01	$13 \times 9 \times 0.78$
151.2	0.12 ± 0.02	$13 \times 9 \times 0.80$
151.6.1	0.36 ± 0.02	$9.5 \times 9.5 \times 0.25$
151.7	0.74 ± 0.05	$13 \times 9.5 \times 0.60$
151.12.1	1.63 ± 0.06	$9.7 \times 9.5 \times 0.25$
151.5.1	2.92 ± 0.11	$9.3 \times 9.3 \times 0.25$
151.10.1	3.49 ± 0.11	$9.5 \times 9.5 \times 0.25$
151.3	15.30 ± 1.34	$13.7 \times 9.5 \times 0.80$
142.3.2 (z-cut)	0.16 ± 0.02	$10 \times 10 \times 1$

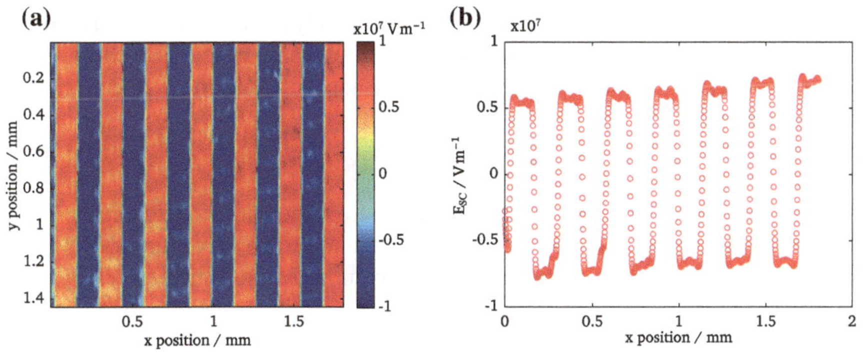

Fig. 4.1 Measurement of field modulation inside a photorefractive sample; crystal *AT1* is illuminated with a binary stripe pattern; a two-dimensional electric field (**a**) pattern is calculated according to Eq. 3.33 and averaged along the constant direction to increase the SNR (**b**)

Since the DHM technique enables single-shot acquisition of phase information, not only the spatial structure and magnitude of the internal electric field can be measured, but also its temporal evolution by taking multiple images of the holographic interference pattern. For each timestep, including the background phase offset at $t = 0$, the two-dimensional phase information can be calculated using the algorithm presented in Sect. 3.6.3.2. After the background phase curvature has been subtracted, the phase retardation is converted into the internal electric field magnitude. Most often, if quantitative data is to be obtained from a one-dimensional modulation, the Signal-to-Noise-Ratio (SNR) can be increased by averaging along the constant direction (in this case the y direction) (Fig. 4.1).

The low photoconductivity that has made $LiNbO_3$ the material of choice for data storage application, restrains its practical use as substrate material in POT. From preliminary measurements, it could be concluded that for an as-grown sample, the writing time to reach an electric field strength of $2\,\mathrm{kV\,mm^{-1}}$, which is far below the saturation of the material, at the low intensity level of $I = 2.7\,\mathrm{mW\,cm^{-2}}$ is approximately 2.5 h. Such time constants, of course, are of little use for practical applications. Three routes exist to significantly reduce the writing time: for not to high intensity levels, the photoconductivity is directly proportional to the incident intensity [4]. This implies that either a stronger laser can be used to increase the overall power or the active area on the crystal can be made smaller, provided that in a particular application not the whole crystal surface is needed. However, both strategies have an ultimate limit in the available laser power or the minimum region that is to be structured. After all, one of the main advantages of POT over direct optical tweezers is their large active area. As a third alternative to faster crystals, it can be exploited that the photoconductivity in iron-doped $LiNbO_3$ samples is proportional to the reduction degree of the material [7]. In the band transport model—modified for the particular case of iron-doped $LiNbO_3$—electrons are excited from filled Fe^{2+}

traps into the conduction band where they preferentially travel along the c-axis to the +c face due to the bulk photovoltaic properties of the material [8]. The photovoltaic current density is therefore proportional to the intensity and the concentration of Fe^{2+} centers [4]:

$$j_{\text{photo}} = \beta_{31,33} I c_{\text{Fe}^{2+}}, \tag{4.1}$$

where $\beta_{31,33}$ is the respective tensor coefficient from the photovoltaic tensor. These free electrons are recaptured by the Fe^{3+} acceptor sites. The maximum photovoltaic field that can be achieved is proportional to the number of redistributed charge carriers, hence to the trapping site concentration:

$$E_{\text{max,pve}} = \frac{\beta_{31,33}\gamma_{\text{r}}}{e\mu s} c_{\text{Fe}^{3+}}, \tag{4.2}$$

where γ_{r} is the recombination rate, e the electron charge, μ the electron mobility and s the ionization cross-section. For cases in which the dark conductivity can be neglected—as it is the case for the intensities used in this work—the photoconductivity can be described as [4]

$$\sigma_{\text{photo}} = \frac{j_{\text{photo}}}{E_{\text{max,pve}}} \propto \frac{c_{\text{Fe}^{2+}}}{c_{\text{Fe}^{3+}}}, \tag{4.3}$$

which means that the photoconductivity rises linearly with the reduction degree, that is, the ratio between filled impurity centers of Fe^{2+} valence state and the empty sites of Fe^{3+} valence state. The straightforward route to faster crystal is hence to increase this reduction degree in iron-doped $LiNbO_3$ samples. In this thesis, it will be investigated, how far crystal samples can be reduced to increase the photoconductivity and still be able to efficiently trap matter on their surface.

4.1.2 Sample Preparation and Reduction Treatment

For the investigation of highly-reduced $LiNbO_3$, several crystal samples were prepared by cooperation partners from the *Ferroelectric Materials Research Group* at the University of Padua in Italy. A crystal boule was grown by the Czochralski technique at a pulling rate of 2 mm h^{-1} [9, 10]. This technique uses a seed crystal with predetermined orientation along the c-axis to start the nucleation from the crystal melt and hence control the growth direction of the boule as well as its crystallographic orientation [11]. In order to introduce additional energy levels in the band gap of the material to enhance its sensitivity in the visible, Fe_2O_3 was added to the melt in a concentration of 0.055 wt%. In the crystal, this corresponds to an iron concentration of c_{iron} of $18.8 \times 10^{18}\text{cm}^{-3}$, which was determined as the highest reasonable iron concentration [4]. After the growth process, the boule is electrically poled at a temperature of 1200 °C to obtain a single domain crystal. This elevated temperature is necessary due to the fact that the coercive field for the inversion of c-axis domains in $LiNbO_3$ at room temperature is approximately 21 kV mm^{-1} [12], which makes it impossible to pole a macroscopic boule using the application of electric fields in air at room temperature. For congruent $LiNbO_3$ with a small lithium deficit, 1200 °C is higher than the Curie temperature [13, 14], above which the material loses its ferroelectric properties. Therefore, the poling process is greatly facilitated and can in fact be accomplished with fields on the order of Vcm^{-1} [15]. After the poling process, the crystal boule is oriented along the three main axes using high-resolution X-ray diffraction. This allows to cut samples of arbitrary orientation, denoted by the direction normal to the surface, for example *y-cut* with the axis of charge transport being in-plane, and subsequently polish them to optical quality in a commercial polishing machine from Logitech.

In the as-grown state without any additional treatment, the iron incorporated into the crystal is mainly in its 3+ valence state, resulting in a low photoconductivity and hence high writing times for internal space-charge fields. To modify the reduction degree, the samples undergo a reduction treatment for different times at 500 °C in a gas mixture of Ar(98 %)+ H_2(2 %) [16]. The ratio between Fe^{2+} and Fe^{3+} can be assessed by spectroscopic measurements. Since, the absorption of the Fe^{2+} band centered around 481 nm is proportional to its concentration, an absorption spectrum is acquired in the range of 330–2000 nm. The transmission through any absorbing sample generally can be described as

$$T_S = \frac{I_T}{I_0} = \gamma e^{-\alpha_S d}, \tag{4.4}$$

where I_T the transmitted intensity, I_0 the incident intensity, α_S the absorption of the sample and d its thickness. γ is a correction factor, which accounts for the Fresnel losses from refractive index jumps. The measured transmittance is related to the internal material transmittance T_i by [17]

$$T = \frac{(1-R)^2 T_i}{1 - R^2 T_i^2}. \tag{4.5}$$

In this equation, $R = ((n_2 - n_1)/(n_2 + n_1))^2$ is the Fresnel reflection coefficient and $n_{1,2}$ are the refractive indices of air and $LiNbO_3$, respectively. The corrected internal transmittance, however, also includes the intrinsic absorption of undoped $LiNbO_3$. Therefore, by normalizing the transmission of the iron-doped sample to the transmission of an undoped pure crystal, it is possible to obtain only the absorption A due to Fe^{2+} centers in the material [17]:

$$A = \ln \frac{T_{\text{crystal}}}{T_S} = (\alpha_S - \alpha_{\text{crystal}})d \tag{4.6}$$

Knowing the respective scattering cross-section of the iron dopants at a particular wavelength and the total iron concentration in the sample, both c_{Fe2+} and the reduction degree r can be calculated via:

$$c_{Fe^{2+}} = \frac{\alpha_S - \alpha_{crystal}}{\sigma_\lambda} \tag{4.7}$$

$$r = \frac{c_{Fe^{2+}}}{c_{total} - c_{Fe^{2+}}} \tag{4.8}$$

The absorption values were acquired at a wavelength of 532 nm, for which the cross-section has been carefully measured in [18] to be $\sigma_{532} = (3.92 \pm 0.08) \times 10^{-18}\text{cm}^{-2}$ and $\alpha_{crystal}$ has been taken as $(0.029\,66 \pm 0.047\,44)\ \text{cm}^{-1}$ (private correspondence to Annamaria Zaltron, University of Padua). This method was used to determine the reduction degree of a commercial crystal, for which the dopant concentration was specified as 0.055 wt% Fe_2O_3, but the reduction degree was unknown. From the absorption $(2.803 \pm 0.040)\ \text{cm}^{-1}$ at 532 nm, the concentration of Fe^{2+} centers could be calculated to be $(7.02 \pm 0.36) \times 10^{17}\text{cm}^{-3}$ and the reduction degree to $r = 0.039 \pm 0.004$, which corresponds to the as-grown state, as it was obtained by the collaboration partners in Padua in crystals that did not undergo any post-treatment. For the experiments regarding the investigation of the reduction degree, only y-cut samples were prepared. However, for some of the applications that will be presented in this thesis in Chap. 6, crystals have been cut in the z-cut geometry. If not stated otherwise, all the crystal were prepared using the above mentioned procedures.

4.1.3 Measurement of Charge Transport Parameters

The most important figure-of-merit for POT is the magnitude of the saturated internal field that can be created. To obtain the saturation value for the internal fields, $LiNbO_3$ samples with different reduction degrees were illuminated with a sinusoidal amplitude pattern and the resulting space-charge field modulation was acquired two-dimensionally by DHM as presented previously. The dynamic measurement of the field was stopped when a steady state was reached. Note that in contrast to other publications, here the full modulation of the field $E_{full} = E_{max} - E_{min}$ was determined. The reason for this presentation is that in the saturated case, there is a strong deviation of the internal field from the assumed sinusoidal intensity pattern which can possible be attributed to the occurrence of higher harmonics of the field components (see also Sect. 4.1.4). In order to account for the asymmetry of the field, its full modulation was chosen as the measurement parameter. However, when comparing the values obtained here to those from other publications, it can be assumed that in publications using the Kogelnik method of diffraction efficiency, the field is assumed to be symmetrical, hence $E_{full} = 2 \cdot E_{SC}$.

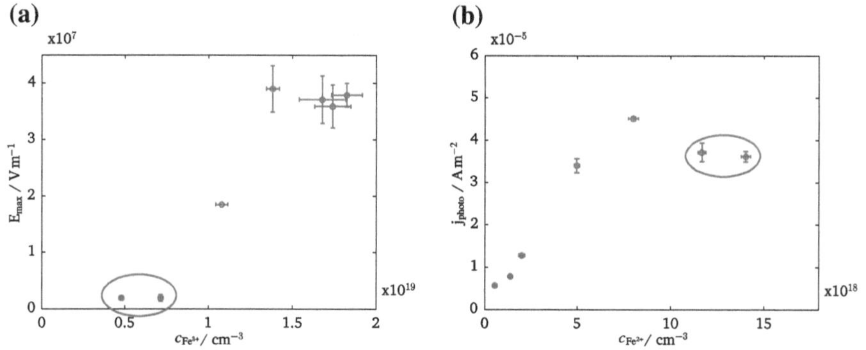

Fig. 4.2 Photorefractive properties of highly reduced $LiNbO_3$ samples; the measurements show a significant deviation from the predictions of the one-center model for crystals with a reduction degree of 1.63 and 2.92, the samples indicated by red circles (also published in [7])

By plotting the maximum value of E_{SC} obtained for each sample as a function of Fe^{3+} content, as presented in Fig. 4.2a, it is clearly visible that there is a strong deviation from the behavior as it would be expected from the theory of the one-center model. For Fe^{3+} concentrations down to 1.1×10^{19}cm^{-3} the field decreases linearly with $c_{Fe^{3+}}$, as reported earlier [4]. For samples with a higher reduction degree than 0.74, we find a sudden decrease of the space-field by more than one order of magnitude. This behavior points at the occurrence of different mechanisms of charge transport, which may deteriorate the internal field. However, by monitoring the decay of the internal field in the non-illuminated state, a significantly increased dark conductivity in these samples could not be verified, which indicates that additional charge transport mechanisms must be present in the illuminated state.

Figure 4.2b shows the measurement of the photovoltaic current density plotted against the Fe^{2+} concentration. In first approximation, this value determines the writing speed of an internal field distribution. Due to the different absorption values and thicknesses of the crystals (cf. Table 4.1), special care was taken that the mean intensity of the green writing laser beam along the crystal thickness was $I = (49 \pm 1)$ mW cm^{-2} for all samples, so that the current density is comparable. A linear increase of this value is clearly seen up to concentrations of $(8.00 \pm 0.54) \times 10^{18}$cm^{-3}, whereas for reduction degrees higher than 0.74 ± 0.05 the value of the current density decreases. As a consequence, the photoconductivity (cf. Fig. 4.3), for which a linear increase with the reduction degree is expected, shows a strong super-linear component for reduction degrees above 0.74. This can be understood when one keeps in mind that the decrease of the electric field can be caused by any kind of increased conductivity, whereas the photovoltaic current density describes only the direct charge transport due to the PVE. Therefore, the overall conductivity can rise substantially although the conductivity due to the bulk photovoltaic effect saturates. In order to get a closer insight into the origin of these previously suggested additional

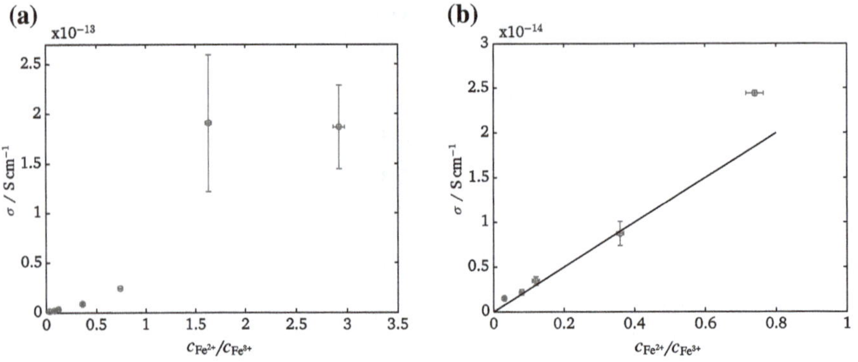

Fig. 4.3 Photoconductivity of the highly reduced $LiNbO_3$ samples as obtained from the photorefractive parameters; (**b**) shows a closeup of the linear increase of the photoconductivity for samples up to a reduction degree of 0.74, whereas there is a superlinear increase for higher reduction degrees, a strong indicator for other charge transport mechanisms (also published in [7])

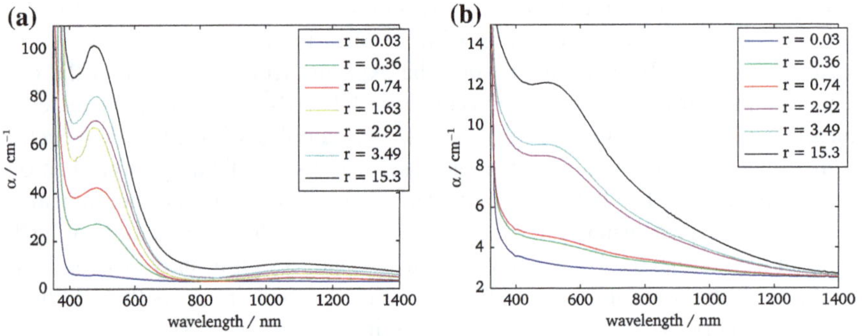

Fig. 4.4 Absorption spectra of highly reduced samples (reduction degree r) and the absorption contribution due to the occurrence of polarons/bipolarons (inset) (also published in [7]). **a** Total absorption, **b** (bi-)polaron absorption

charge transport processes, the obtained absorption spectra, as shown in Fig. 4.4 were fitted as a combination of two absorption processes; the absorption due to the Fe^{2+} bands centered at 466 and 1100 nm and the bipolaron/polaron bands centered at about 495 and 755 nm, respectively. By this approach is was possible to remove the absorption due to iron, hence to amplify the contribution due to the absorption of polarons and bipolarons, which show a strong absorption in the highly reduced samples and could be one possible reason for the superlinear increase of the photoconductivity [19].

Another possible yet not fully tested explanation is incorporation of hydrogen during the reduction treatment. It has been shown in earlier publications by the cooperation partners from Padua that a mixture of 98 % Ar + 2 % H_2 promotes the reduction of iron without causing any problems, such as an unexpected decrease of saturation field [16, 20]. However, to obtain reduction degrees of more than 1, the reduction time is on the order of 10 h, during which an increased amount of hydrogen could have been introduced into the sample. The hydrogen ions' mobility and hence their contribution to the total conductivity rises significantly with temperature [21]. From thermal simulations carried out at the University of Padua, it could be calculated that the sample temperature increases by 35 K [7]. This could be a second explanation for the observation that obviously an increased overall conductivity reduces the saturation field, but no increased dark conductivity in the absence of laser illumination could be observed.

4.1.4 Electric Field Structure at High Modulation Depths

An interesting result from the DHM measurement of the internal space-charge field is that the magnitude of the field is approximately a factor of 2–3 larger than the highest values ever obtained by the Kogelnik diffraction method [8]. The possibility of an erroneous measurement or faulty phase unwrapping procedure could be ruled out by repeating the experiments, comparing the results for several crystals and manual inspection of the unwrapped phase images during the DHM measurement process. To make both methods directly comparable, the saturation field for crystal sample *AT1* has been measured with the Kogelnik method, in addition to the value derived by DHM in the previous section. Figure 4.5 shows for comparison the temporal electric field evolution for both techniques. In the case of the Kogelnik method, the evolution follows nicely a mono-exponential law, that is frequently assumed for the temporal evolution of the internal grating formation in photorefractive materials (cf. [22] and also Eq. 3.6). On the contrary, the behavior of the field magnitude in the case of DHM differs significantly from this law.

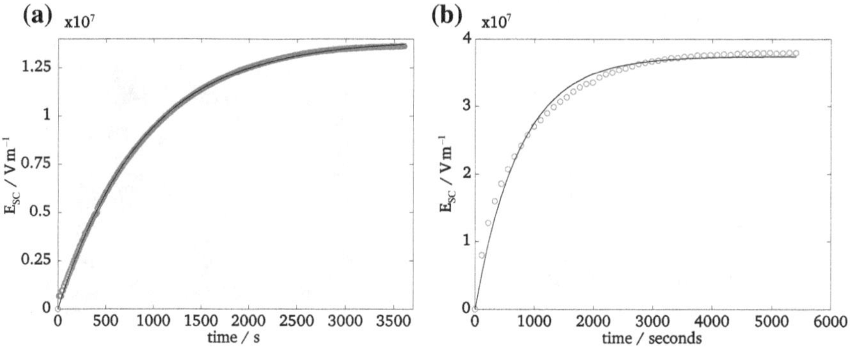

Fig. 4.5 Comparison between internal electric field magnitude in an as-grown $LiNbO_3$ crystal (AT1) obtained indirectly by measuring the diffraction efficiency and directly by DHM measurement. **a** Kogelnik method, **b** DHM method

A possible explanation might be found in the derivation of the solution of Kukhtarev's equations. In order for the harmonic ansatz to be valid, the modulation of the light pattern that induces the refractive index grating was assumed to be very low, hence $m \ll 1$. The approximation of purely sinusoidal refractive index structure was also assumed by Kogelnik in his derivation of the diffraction efficiency [23]. However, for both the techniques, the modulation of the illumination is almost equal to 1, regardless if the field is induced by interference (as for the Kogelnik method) or by an amplitude modulated image from an ASLM. It has been noted in theoretical models of the photorefractive grating formation that for modulation depths $m \approx 1$, higher harmonics with different time constants as compared to the fundamental harmonic should play a significant role and contribute to the internal field [24, 25]. Figure 4.6 focuses on the electric field structure rather than its magnitude. While after 220 s, the internal electric fields still mimics the incident sinusoidal intensity modulation, its structure has changed significantly after 990 s of illumination with the same pattern. The field has been narrowed down to very thin high field areas, while the areas of low electric field become broader. The theoretical prediction of higher harmonics contributing to the internal electric field can be clearly observed in Fig. 4.6d, which shows the temporal evolution of the Fourier components. It can be clearly seen that after 220 s or 0.06 h, additional higher harmonic components start

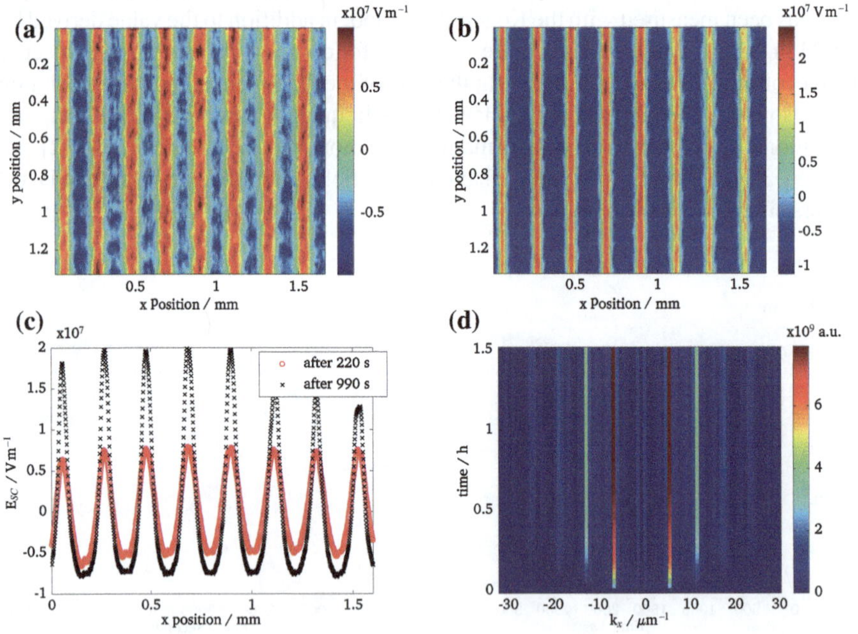

Fig. 4.6 Electric field form at high modulation depths; crystal *AT1* was illuminated by a sinusoidal light pattern (period $\Lambda = 211\,\mu\text{m}$) at an intensity level of $0.68\,\text{Wcm}^{-2}$; while after 220 s (**a**), the electric is a replica of the sinusoidal intensity pattern, its form is considerably changed after 990 s (**b**). Averaged electric field structure (**c**) and temporal evolution in Fourier space (**d**), indicating that higher harmonics start to contribute after 220 s

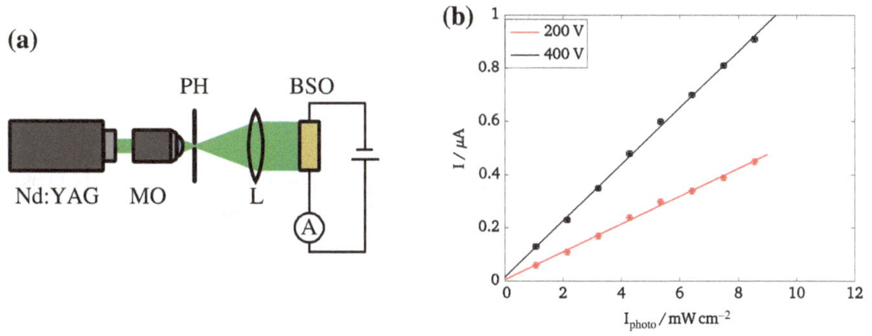

Fig. 4.7 Direct measurement of the photoconductivity for BSO samples. **a** Setup, **b** results for the photocurrent

to evolve. Before the contribution of the additional terms, the total field modulation is $E_{SC} \approx \pm(1.24 \pm 0.20) \times 10^7$ V m^{-1}, hence comparable to the value that could be obtained for the same crystal employing the Kogelnik method. These observations, which to the best of the author's knowledge represent the first direct experimental observation of higher harmonics with DHM, could serve as a hint to the fact that the internal fields in $LiNbO_3$ can be much higher than previously determined but that they cannot be probed using the Kogelnik technique because it only describes diffraction from the fundamental grating and not from higher harmonics.

4.2 Bismuth Silicon Oxide (BSO)

4.2.1 Photoconductivity and Real-Time Induction of Space-Charge Fields

Bismuth silicon oxide—as compared to $LiNbO_3$—is a photorefractive material that has a photoconductivity which is orders of magnitude larger. The consequential high recording speed for internal space-charge field has inspired multiple holographic applications such as phase-conjugation for technical inspection [26] or photorefractive image correlation [27]. Since its photovoltaic field can be neglected [3], this material will serve as a model system for drift-dominated photorefractive media. BSO belongs to the family of sillenites, which all feature similar properties, such as high photoconductivity, moderate electro-optical coefficient and significant optical activity [28]. Its significantly higher photoconductivity allows direct conductivity measurements. To this end, the crystal is sandwiched between two electrodes to which a voltage is applied. The sample is fully illuminated with an expanded 532 nm beam and the current through the sample is measured with a pico-amperemeter. The setup as well as the results of the current measurements are illustrated in Fig. 4.7.

The total conductivity can be derived as the slope of the conductivity with increasing light intensity I_{ph}

$$I = (\sigma_{\mathrm{ph}} I_{\mathrm{ph}} + \sigma_{\mathrm{dark}})U \tag{4.9}$$

where the photoconductivity σ_{ph} is calculated from the crystal length l, its cross-section A, the applied voltage U, the electrical current I, and the slope of the fit $m = I/I_{\mathrm{ph}}$ in Fig. 4.7 as

$$\sigma_{\mathrm{ph}} = \frac{l}{UA}\frac{I}{I_{\mathrm{ph}}} = \frac{l}{UA}m. \tag{4.10}$$

Note that the incident light intensity varies over the crystal length due to absorption. The mean intensity is calculated as $I_{\mathrm{ph}} = (I_0/(\alpha d))(1-\exp(-\alpha d))$, with the incident intensity I_0. However, since the absorption of BSO at 532 nm is very low–α was measured to be approximately 0.8 cm^{-1}—the correction is below 10 %. The same experiment has been repeated twice, with applied voltages of 200 and 400 V. The mean value of both measurements for a crystal of dimensions $l = 10.2$ mm and $A = 15\ \mathrm{mm}^2$ is $\sigma_{\mathrm{ph}} \approx (1.79 \pm 0.01) \times 10^{-9}\mathrm{S\ cm\ W}^{-1}$.

4.2.2 Temporal Electric Field Response of BSO

The high photoconductivity[1] allows the speed of recording for internal fields to approach real-time. However, an important challenge is the fact that due to the high photoconductivity, the voltage drop in the illuminated region is significantly decreased. Figure 4.8 shows a typical illumination scenario for POT, in fact for all kinds of optoelectronic tweezers: an external voltage is applied over the whole active surface and only the part of the crystal where trapping of particles is desired is illuminated with a modulated pattern of any kind. The predicament for the geometry of photorefractive tweezers that originates from this geometry is illustrated by the respective equivalent circuit in Fig. 4.8b. Since from Ohm's law it is known that the voltage drop is proportional to the resistance of the crystal, in the illuminated state most of the voltage drop is in the non-illuminated parts of the crystal, whereas the electric field is pushed out of the interesting area, yielding a very low modulation for E_{SC}, hence little to no dielectrophoretic forces.

As a possible solution to the problem illustrated in Fig. 4.8, Eason and Vainos suggested to use a separate laser beam to illuminate the portions of the crystal where no voltage drop is desired [29]. Intriguing as this scheme might be, it requires the exact adjustment of the auxiliary laser to the size of the trapping pattern imaged onto the crystal. In photorefractive optics, where BSO was frequently used for image processing and phase conjugation [26, 27], the diminished internal field was described phenomenologically with the built-up of screening charges that drift out of the

[1] Results partly obtained in collaboration with Hannes Futterlieb during his Bachelor thesis.

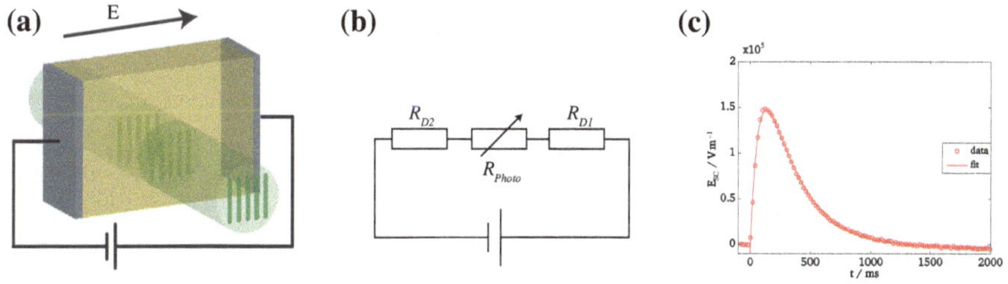

Fig. 4.8 Typical illumination setup for photorefractive opto-electronic tweezers: The crystal is illuminated in an active region, lowering the voltage drop and hence the electric field in this region. **a** Typical setup, **b** equivalent circuit, **c** temporal evolution of screening fields

illuminated region and shield the illuminated region against the external field [30]. To investigate this behavior, a BSO crystal in $[\bar{1}10]$ geometry was equipped with external electrodes, integrated into the DHM setup and illuminated with a sinusoidally modulated light pattern. Note that in this case a separate setup was used, where the reflective HED 6001 modulator was replaced by a transmissive LC 2002 modulator (Holoeye, Berlin) with reduced resolution of 800×600 pixels but otherwise comparable figures-of-merit. The quantitative values for the phase modulation were again derived by averaging the sinusoidal pattern along the constant direction and the full modulation $E_{SC} = E_{max} - E_{min}$ was plotted. To supply a sufficiently high external field, the crystal electrodes were connected to a bipolar amplifier (*609E-6*, Trek Inc., Medina, New York) capable of providing up to ± 4 kV at a bandwidth of 6 kHz.

The results from a time-dependent investigation of the internal field modulation for two different incident intensity levels can be seen in Fig. 4.9. The laser light pattern, which does not cover the whole crystal surface, is switched on at $t = 0$ s. After an initially very steep increase of the field modulation, it is then decreased by the build-up of the screening field, which totally screens the internal field in the steady-state after approximately 1500 ms for $I = 19\ \mu\text{Wcm}^{-2}$. The temporal evolution of E_{SC} can be described as a superposition of the two components from the desired sinusoidal modulation and the detrimental screening charges:

$$E_{SC}(t) = E_{int}(1 - e^{-\frac{t}{\tau_{int}}}) - E_{screen}(1 - e^{-\frac{t}{\tau_{screen}}}) , \tag{4.11}$$

with $E_{int} = (3.54 \pm 0.15)$ kV cm^{-1}, $E_{screen} = (3.58 \pm 0.14)$ kV cm^{-1}, $\tau_{int} = (74.52 \pm 2.80)$ ms and $\tau_{screen} = (281.82 \pm 7.99)$ ms. From the fit parameters, it can be seen that while both contributions have a comparable field magnitude which cancels out after approximately 1.5 s, their time constants differ significantly, which can most likely be attributed to the fact that the screening requires a larger drift distance for electrons. In order to circumvent the problems associated with the built-up of screening charges and as a very positive side effect further enhance the electric field, it has been suggested to to apply an alternating electric field instead of a DC field [31–33]. Provided that the period of the external electric field is significantly

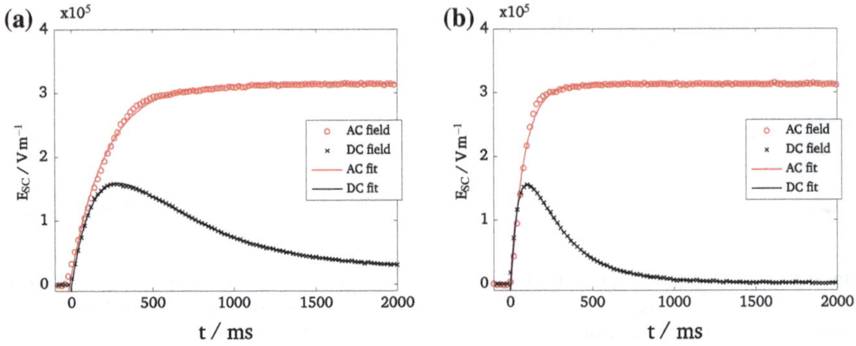

Fig. 4.9 AC/DC response of the field modulation in BSO for two different intensities: In the presence of external DC fields, screening charges effectively screen the field from the active region in the steady state, whereas in the case of an applied AC field ($f = 100\,\text{Hz}$), the occurrence of spatial screening charges is suppressed. **a** Temporal evolution for $19\,\mu\text{W}\,\text{cm}^{-2}$, **b** temporal evolution for $52\,\mu\text{W}\,\text{cm}^{-2}$

shorter than the grating formation time, there is no preferential direction of electron movement, hence suppressing the occurrence of screening charges through large scale drift. On the other hand, the external field period should be sufficiently large as to be considered constant over the carrier lifetime τ_r. In this scenario, the electrons efficiently experience a constant electric field over their lifetime. It can be shown that the criterion

$$\tau_{\text{grating}} \ll T_{\text{field}} \ll \tau_r, \tag{4.12}$$

which seems to restrict the external field frequencies, is a rather weak restriction, since typical grating formation times are in the range of 100 ms or less (cf. Fig. 4.8c), whereas the carrier lifetime is in the range of ns [28]. Therefore, external field frequencies in the very range of 10 Hz–10 MHz are sufficient and suitable to suppress the screening fields.

Several authors have undertaken the task of solving Kukhtarev's equation in the presence of a temporally varying external field [31–34]:

$$E(t) = \hat{E}q(t), \tag{4.13}$$

where $\hat{E}$ is the amplitude of the external field and $q(t)$ is a periodic function describing its actual form. In this case the expression of the internal space charge field E_{SC} changes to [32, 34]

$$E_{\text{SC}} = \frac{1}{2}\left\{\frac{iE_d + E_A^{'}}{1 + (iE_d + E_A^{'})/E_q}\right\}, \tag{4.14}$$

with

$$E_A^{'} = i\eta\hat{E} = i\left\{\frac{1}{I_0(P)} - \frac{1}{P}\right\}\hat{E}, \tag{4.15}$$

where

$$I_0(P) = \frac{1}{T} \int_{-T/2}^{T/2} \frac{P}{1 - iPq(t)} dt \tag{4.16}$$

is an integral over the waveform and $P = (\mu\tau_r E_0 K)/(1 + D\tau_r K^2))$ is a parameter defined by the experimental parameters and material properties, such as the charge carrier mobility μ, the charge carrier lifetime τ_r, the diffusion coefficient D and the grating vector K. Formally, Eq. 4.15 shows that for the case of an externally applied alternating field, the amplitude of the external field which limits the internal field strength for the DC case, is modified by an enhancement factor η. For the typical sinusoidal and square waveforms, the integrals can be easily solved. For both waveforms, the integral can be made symmetric by an appropriate substitution and be solved for $0 \leq t \leq T/2$, arriving at [32]:

$$\begin{aligned} I_0(P) &= \frac{P}{1 + P^2} \\ \rightarrow \eta &= P \end{aligned} \tag{4.17}$$

for the square waveform and

$$\begin{aligned} I_0(P) &= \frac{P}{\sqrt{1 + P^2}} \\ \rightarrow \eta &= \frac{\sqrt{P^2 + 1} - 1}{P} \end{aligned} \tag{4.18}$$

for the sinusoidal waveform. For small P values, the enhancement factor η reduces to $P/2$, for large values of P, $\eta \approx 1$. In both cases, it can be concluded that the field enhancement is larger for the square wave than for the sinusoidal one. Several important aspects can be seen when the solution of the time-averaging approach is discussed: The resulting effective field $\hat{E}$ is purely imaginary, which means that there is a $\pi/2$ phase shift between the inducing light pattern and the resulting electric field, resembling the occurrence of diffusion fields but without the strong dependence on the grating period. This fact has been previously employed in photorefractive optics to demonstrate efficient two-wave mixing inside photorefractive BSO crystals, for which a phase-shift of $\pi/2$ is a prerequisite since otherwise there would be no energy transfer between interfering beams [35].

The knowledge about the application of an external field to suppress screening charges can be used to sustain and even enhance the internal field modulation, which is highly beneficial for the task of dielectrophoretic trapping. Figure 4.9 shows the internal electric field modulation strength for both the DC and AC case at two different intensity levels. In all cases, the externally applied field was of square form, had a magnitude of 360 V mm^{-1} and a frequency of 100 Hz. The incident sinusoidal intensity had a period of $\Lambda = 280\,\mu$m with a modulation of $m = 1/2$. It is clearly obvious that in both cases the criterion of $T_{\text{field}} \ll \tau_{\text{grating}}$ is fulfilled, the occurrence of screening charges is suppressed and a constant electric field modulation is obtained. As for all photorefractive materials, the grating formation time is inversely proportional to the incident intensity, so that a steeper rise time is expected for higher intensities (cf. Fig. 4.9b). Since the field magnitude E_{max} is proportional to the modulation degree of the pattern m, the enhancement factor in this case is 0.92. Please note that due to the optical activity of the material, this enhancement factor is only a lower estimate. For crystals of 3 mm thickness, as the one used in this particular experiment, it is expected that the real linear birefringence is much larger than the measured one (cf. also Sect. 3.5.1 and [36]). Nevertheless, the general conclusion remains unchanged, since there is only a proportionality factor between real field and measured field.

4.2.3 AC Response of Internal Space-Charge Fields in BSO

The theoretical findings about the internal field formation in the presence of an alternating external voltage can be investigated in more detail if the parameter space is expanded. To verify the non-resonant grating enhancement condition $T_{\text{grating}} \ll \tau_{\text{grating}}$, the internal field modulation is measured while varying the AC field frequency. Figure 4.10 shows the total field modulation in the saturation case plotted against the field frequency for different intensity levels. For each level, both square and sinusoidal waveforms were applied. The results corroborate the previous findings, since it can be seen that for a DC applied field, the stationary field is significantly decreased by the screening charges. The field modulation increases as the AC field frequency is increased until it reaches a plateau of saturation, where the

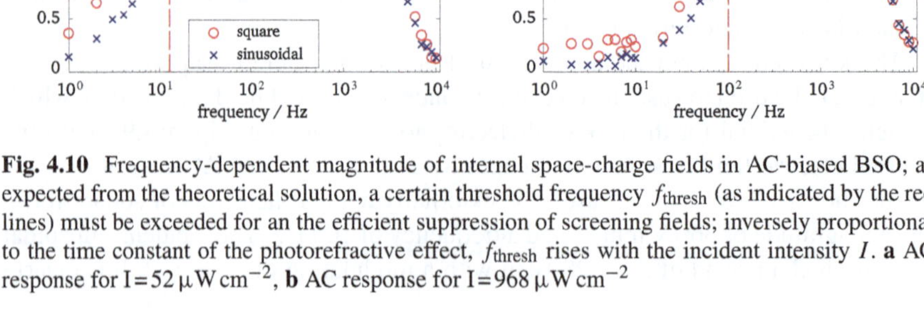

Fig. 4.10 Frequency-dependent magnitude of internal space-charge fields in AC-biased BSO; as expected from the theoretical solution, a certain threshold frequency f_{thresh} (as indicated by the red lines) must be exceeded for an the efficient suppression of screening fields; inversely proportional to the time constant of the photorefractive effect, f_{thresh} rises with the incident intensity I. **a** AC response for $I = 52\,\mu\text{W}\,\text{cm}^{-2}$, **b** AC response for $I = 968\,\mu\text{W}\,\text{cm}^{-2}$

maximum enhancement factor occurs for a large range of frequencies. For very high frequencies, the grating strength decays to zero, although these frequencies are still significantly smaller than the inverse of the free carrier lifetime. The explanation for this effect is that for higher frequencies, the voltage amplifier reaches its bandwidth limit. Despite the fact that the manual specifies a large-signal bandwidth of 6 kHz, a significant drop of the efficiency can already be seen at $f = 2\,\text{kHz}$. The effect is stronger for an external field of square waveform, since many higher harmonic terms are needed in this case and even a slight deviation from the perfect square form can significantly reduce the enhancement efficiency. This is the reason why for practical applications, Walsh et al. suggest the use of sinusoidal waveforms in spite of their lower efficiency, since these are easier to realize using driven resonant circuits [32].

The frequency at which the plateau is reached increases with the incident intensity, which is not surprising since the grating formation time scales inversely to the intensity, so in order to satisfy Eq. 4.12, the field frequency must be increased with increasing illumination intensity. To arrive at a quantitative guideline for further experiments, a threshold frequency for each intensity level is defined as the point, where 80 % of the full modulation of E_{int} has been reached. This threshold frequency shows a linear increase at low light intensities and saturates at higher values for I (cf. Fig. 4.11). Please note that this observation only refers to the suppression of screening fields and not to the bulk photoconductivity, where, in a different study, no saturation was observed even for intensities of 1000 mW cm^{-2} [37]. Summing up the findings for BSO, the non-resonant nature of the AC-field enhancement condition opens up a very large frequency range which can be used to suppress the occurrence of screening charges and to enhance the internal electric field as indicated in Fig. 4.11. Ideally, the frequency of the external field is chosen in such a way that the condition $\tau_{\text{grating}} \ll T_{\text{field}} \ll \tau_r$ is fulfilled for all intensities that are expected in the experiment.

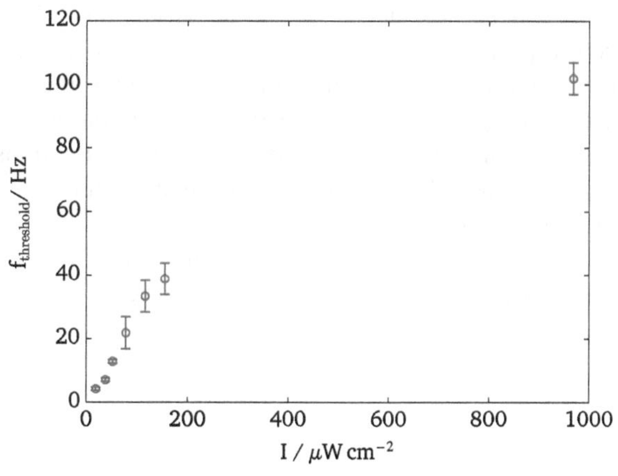

Fig. 4.11 Threshold frequency plotted against the incident intensity I; with increasing intensity, the external frequency must be appropriately adjusted to fulfill Eq. 4.12

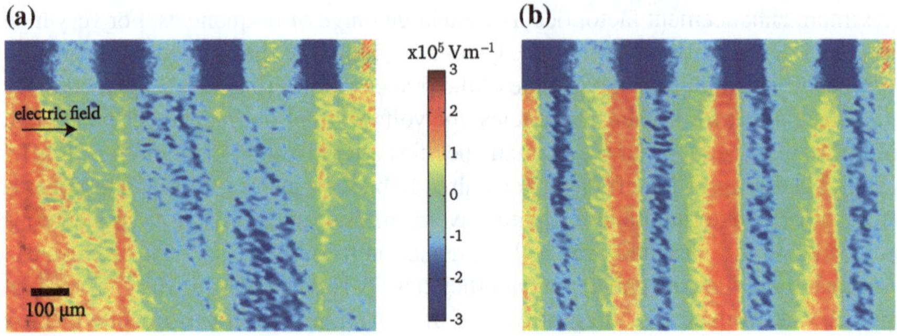

Fig. 4.12 Field enhancement in BSO: **a** for a DC bias, the internal field remains in-phase, but is of low modulation; **b** an AC-field shifts the field by $\pi/2$ and even enhances the internal field above the external value of 2×10^5 V m^{-1}; *top* insets show incident intensity pattern (also published in [38])

4.2.4 Electric Field Structure and Phase-Shift Inside BSO

The two-dimensional measurement capabilities of DHM allow the correlation of the incident intensity pattern with the resulting electric field. The two cases of a DC and an AC external voltage have been compared under the incidence of a sinusoidal illuminating pattern in Fig. 4.12. All the previous findings about the field modulation can be nicely visualized in this example: for a DC field, the modulation of E_{SC} is greatly diminished, but still in-phase with the intensity pattern. In the case of an AC bias, however, the field modulation is an almost exact replica of the illumination pattern, yet phase-shifted by $\varphi = \pi/2$ as indicated by the imaginary factor in Eq. 4.15 [38]. Over and above, the measured internal field is amplified over the external value, as suggested by the previous theoretical models. Note that in this particular experiment, a BSO crystal of 1 mm thickness was used, so that the deviation of the measured space-charge field due to optical activity is much less than in the frequency scans before [36], which is the reason why for thin crystals, a real enhancement can be measured. To the best of the author's knowledge, the findings presented in this section and [38] represent the first spatial examination of the internal fields in BSO using DHM and are direct proof of the theoretical assumptions by Stepanov and Petrov [31] as well many other authors [32, 34] that the internal field in AC-biased crystals is indeed $\pi/2$ phase-shifted with respect to the illuminating pattern. Of course, this finding has already been indirectly inferred from the fact that a very large increase of the two-wave mixing gain for AC-biased crystals could be reported [32, 39].

References

1. B. Matthias, J. Remeike, Ferroelectricity in the Ilmenite Structure. Phys. Rev. **76**(12), 1886–1887 (1949)
2. A. Ashkin, G. Boyd, J. Dziedzic, R. Smith et al., Optically-induced Refractive Index Inhomogeneities in $LiNbO_3$ and $LiTaO_3$. Appl. Phys. Lett. **9**, 72–74 (1966)
3. B. Sturman, V. Fridkin, *The Photovoltaic and Photorefractive Effects in Noncentrosymmetric Materials* (Gordon & Breach Science Publishers, Philadelphia, 1992)
4. K. Peithmann, A. Wiebrock, K. Buse, Photorefractive properties of highly-doped lithium niobate crystals in the visible and near-infrared. Appl. Phys. B Lasers Opt. **68**(5), 777–784 (1999)
5. J. Heanue, M. Bashaw, L. Hesselink, Volume holographic storage and retrieval of digital data. Science **265**(5173), 749–752 (1994)
6. K. Buse, A. Adibi, D. Psaltis, Non-volatile holographic storage in doubly doped lithium niobate crystals. Nature **393**(6686), 665–668 (1998)
7. M. Esseling, A. Zaltron, N. Argiolas, G. Nava et al., Highly reduced iron-doped lithium niobate for opto-electronic tweezers. Appl. Phys. B Lasers Opt. **113**(2), 191–197 (2013)
8. K. Buse, J. Imbrock, E. Krätzig, K. Peithmann, *Photorefractive Materials and their Application, chapter Photorefractive Effects in*$LiNbO_3$ *and* $LiTaO_3$ (Springer, Berlin, 2007), pp. 83–126
9. R. Byer, J. Young, R. Feigelso, Growth of high-quality $LiNbO_3$ crystals from congruent melt. J. Appl. Phys. **41**(6), 2320 (1970)
10. W. Phillips, D. Staebler, Control of Fe^{2+} concentration in iron-doped lithium-niobate. J. Electron. Mater. **3**(2), 601–617 (1974)
11. A. Ballman, Growth of piezoelectric and ferroelectric materials by Czochralski technique. J. Am. Ceram. Soc. **48**(2), 112 (1965)
12. V. Gopalan, T. Mitchell, Y. Furukawa, K. Kitamura, The role of nonstoichiometry in 180 degrees domain switching of $LiNbO_3$ crystals. Appl. Phys. Lett. **72**(16), 1981–1983 (1998)
13. J. Carruthers, G. Peterson, M. Grasso, P. Bridenba, Nonstoichiometry and crystal growth of lithium niobate. J. Appl. Phys. **42**(5), 1846–2000 (1971)
14. M. Palatnikov, N. Sidorov, V. Skiba, D. Makarov et al., Effects of nonstoichiometry and doping on the Curie temperature and defect structure of lithium niobate. Inorg. Mater. **36**(5), 489–493 (2000)
15. K. Nassau, H. Levinste, G. Loiacono, Ferroelectric lithium niobate: 2. Preparation of single domain crystals. J. Phys. Chem. Solids **27**(6–7), 989–1000 (1966)
16. M.V. Ciampolillo, A. Zaltron, M. Bazzan, N. Argiolas et al., Iron doping of lithium niobate by thermal diffusion from thin film: study of the treatment effect. Appl. Phys. A Mater. Sci. Proc. **104**(1), 453–460 (2011)
17. M.V. Ciampolillo, A. Zaltron, M. Bazzan, N. Argiolas et al., Quantification of iron (Fe) in lithium niobate by optical absorption. Appl. Spectrosc. **65**(2), 216–220 (2011)
18. D. Berben, K. Buse, S. Wevering, P. Herth et al., Lifetime of small polarons in iron-doped lithium-niobate crystals. J. Appl. Phys. **87**(3), 1034–1041 (2000)
19. K. Peithmann, N. Korneev, M. Flaspohler, K. Buse et al., Investigation of small polarons in reduced iron-doped lithium-niobate crystals by non-steady-state photocurrent techniques. Physica Status Solidi A Appl. Res. **178**(1), R1–R3 (2000)
20. M.V. Ciampolillo, A. Zaltron, M. Bazzan, N. Argiolas et al., Lithium niobate crystals doped with iron by thermal diffusion: Relation between lattice deformation and reduction degree. J. Appl. Phys. **107**(8) (2010)
21. A. Yariv, S. Orlov, G. Rakuljic, Holographic storage dynamics in lithium niobate: theory and experiment. J. Opt. Soc. Am. B Opt. Phys. **13**(11), 2513–2523 (1996)
22. P. Yeh, *Introduction to Photorefractive Nonlinear Optics* (Wiley Interscience, New York, 1993)
23. H. Kogelnik, Coupled wave theory for thick hologram gratings. Bell Syst. Tech. J. **48**(9), 2909–2947 (1969)
24. E. Serrano, V. Lopez, M. Carrascosa, F. Agullolopez, Steady-state photorefractive gratings in $LiNbO_3$ for strong light-modulation depths. IEEE J. Quantum Electron. **30**, 875–880 (1994)

25. E. Serrano, M. Carrascosa, F. AgulloLopez, Analytical and numerical study of photorefractive kinetics at high modulation depths. J. Opt. Soc. Am. B Opt. Phys. **13**(11), 2587–2594 (1996)
26. A. Marrakchi, J. Huignard, J. Herriau, Application of phase conjugation in Bi12SiO20 crystals to mode-pattern visualization of diffuse vibrating structures. Opt. Commun. **34**(1), 15–18 (1980)
27. H. Rajbenbach, S. Bann, P. Refregier, P. Joffre et al., Compact photorefractive correlator for robotic applications. Appl. Opt. **31**(26), 5666–5674 (1992)
28. M. Petrov, V. Bryksin, *Photorefractive Materials and their Applications 2: Materials, chapter Space-Charge Waves in Sillenites: Rectification and Second-Harmonic Generation* (Springer, Berlin, 2007) pp. 285–325
29. R. Eason, N. Vainos, Photoconductive enhancement of degenerate 4-wave mixing reflectivity in BSO. J. Mod. Opt. **35**, 491–503 (1988)
30. J. Wilde, L. Hesselink, S. McCahon, M. Klein et al., Measurement of electrooptic and electrogyratory effects in Bi12TiO20. J. Appl. Phys. **67**(5), 2245–2252 (1990)
31. S. Stepanov, M. Petrov, Efficient unstationary holographic recording in photorefractive crystals under an external alternating electric-field. Opt. Commun. **53**(5), 292–295 (1985)
32. K. Walsh, A. Powell, C. Stace, T. Hall, Techniques for the enhancement of space-charge fields in photorefractive materials. J. Opt. Soc. Am. B -Opt. Phys. **7**(3), 288–303 (1990)
33. A. Grunnet-Jepsen, C. Kwak, I. Richter, L. Solymar, Fundamental space-charge fields for applied alternating electric fields in photorefractive materials. J. Opt. Soc. Am. B Opt. Phys. **11**(1), 124–131 (1994)
34. C. Stace, A. Powell, K. Walsh, T. Hall, Coupling modulation in photorefractive materials by applying AC electric-fields. Opt. Commun. **70**(6), 509–514 (1989)
35. D. Staebler, J. Amodei, Coupled-wave analysis of holographic storage in $LiNbO_3$. J. Appl. Phys. **43**(3), 1042 (1972)
36. P. Pellatfinnet, Measurement of the electro-optic coefficient of BSO crystals. Opt. Commun. **50**(5), 275–280 (1984)
37. H. Vogt, K. Buse, H. Hesse, E. Kratzig et al., Growth and holographic characterization of nonstoichiometric sillenite-type crystals. J. Appl. Phys. **90**, 3167–3173 (2001)
38. M. Esseling, S. Glaesener, F. Volonteri, C. Denz, Opto-electric particle manipulation on a bismuth silicon oxide crystal. Appl. Phys. Lett. **100**(16), 161903 (2012)
39. J. Kumar, G. Albanese, W. Steuer, M. Ziari, Enhanced 2-beam mixing gain in photorefractive GaAs using alternating electric fields. Opt. Lett. **12**(2), 120–122 (1987)

Chapter 5
Optically-Induced Dielectrophoretic Particle Trapping

After having assessed the relevant properties of the photorefractive substrates, confirming that indeed highly modulated external fields can be created optically in these media, the crystals can be employed for actual dielectrophoretic particle trapping. As mentioned previously, for all the samples under investigations, the time constants for grating formation are on the order of either seconds, as it is the case for $LiNbO_3$, or 100 ms as in the case of BSO. Therefore, it can be concluded that the low-frequency case of the Clausius-Mossotti factor always applies. This means that in order for particles to be attracted to regions of high field intensity, the Clausius-Mossotti factor must be positive, i.e. that the particles' conductivity σ_p has to be larger than that of the surrounding medium. For many of the particles under investigation, such as graphite [1], glassy carbon spheres [2–4] or Zeolyte L nanocontainers [5], this is the case for both air or insulating alcanes as the surrounding media. The choice of an appropriate suspension medium is a challenge that will be discussed in Sect. 5.4. Naturally, the inverse case of particles being rejected high field intensities is also possible and has also been used experimentally [6, 7]. However, the repulsion of particles is of best use in continuous flow systems, where negative DEP forces sort particles based on their size. Furthermore, since particles are pushed away from the substrate, repulsive forces can hardly be quantified in static trapping experiments, which is the reason why for the following demonstration experiments only positive forces are relevant. Negative dielectrophoretic forces will be employed for the advanced optofluidic applications in Chap. 6.

5.1 Bismuth Silicon Oxide (BSO)

As shown in the previous chapter, internal space-charge fields in BSO occur almost instantly, which facilitates very fast optical trapping scenarios. To demonstrate this, a BSO crystal with a length of 10 mm is equipped with liquid silver for better electrical contact and held between copper spring electrodes. This arrangement is integrated into a custom-built microscope, where an infrared LED is used for observation of

M. Esseling, *Photorefractive Optoelectronic Tweezers and Their Applications*,
Springer Theses, DOI 10.1007/978-3-319-09318-5_5

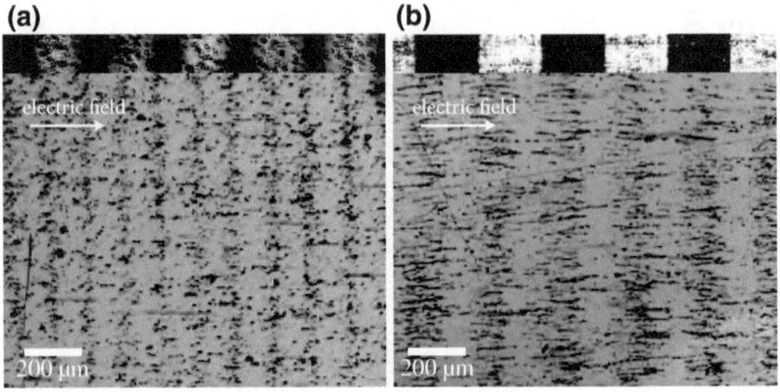

Fig. 5.1 Dielectrophoretic trapping on AC-biased BSO at $f = 100\,\text{Hz}$; sinusoidal and binary patterns show a pronounced doubling of the particle period, a characteristic feature for DEP; trapped particles and the incident light pattern are $\pi/2$ phase-shifted (also published in [8]). **a** sinusoidal pattern, AC voltage, **b** binary pattern, AC voltage

the sample while it is illuminated from below with light patterns generated by an ASLM at $\lambda = 532\,\text{nm}$ as described previously. Since the BSO possesses a negligible absorption in the infrared, the photoconductivity is not substantially affected by this homogeneous illumination but only by the patterned light. Due to the very short writing time for the internal fields, the suspension of particles can be added directly after the writing laser light is switched on. As test particles for the dielectrophoretic manipulation, we use a suspension of homemade graphite flakes ($\phi 6 - 20\,\mu\text{m}$) suspended in tetradecane, an insulating higher alcane oil, so that the assumption of a Clausius-Mossotti factor of 1 is justified. Nevertheless, in accordance with the previous measurements of the internal electric field, little to no particle alignment could be observed in the case of a DC-biased crystal (cf. Fig. 5.2). This can be readily explained with the fact that screening charges have already built up so that the voltage drop in the illuminated region is close to zero. The illumination intensity in this experiment was $260\,\mu\text{W}\,\text{cm}^{-2}$, so it can be concluded from the previous measurements shown in Fig. 4.9 that the internal field is screened after approximately 250 ms, long before particles reach the vicinity of the crystal surface.
On the contrary, if an AC external field with a frequency above the threshold is applied (cf. Fig. 4.11), dielectrophoretic particle trapping can be observed in Fig. 5.1 with its typical characteristic of a frequency-doubling of the particle frequency with respect to the original illumination pattern. Correlating the trapped pattern of particles with the original pattern, the $\pi/2$ phase shift can be also seen in Fig. 5.1: From dielectrophoretic theory, particles are expected to be dragged into regions of high field intensity. For an in-phase sinusoidal grating, these regions would be at the points of highest and lowest intensity. However, due to the phase-shift between pattern and internal field, the trapping positions are also shifted by $\pi/2$, in accordance with the previous determination of the internal field structure in Fig. 4.12. Note that the same

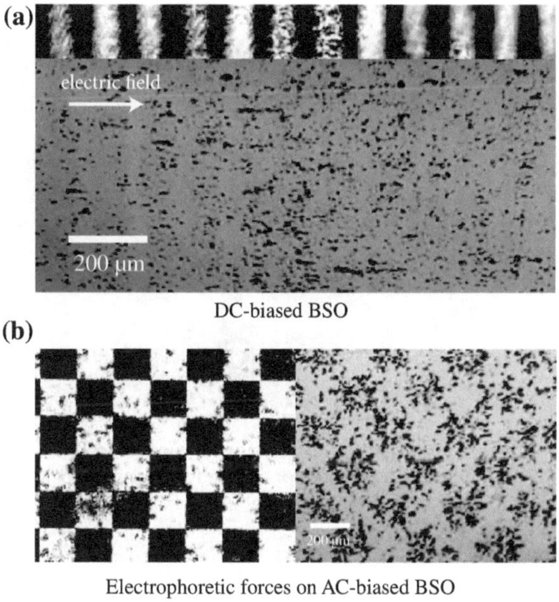

Fig. 5.2 Electrokinetic trapping on BSO: **a** in accordance with previous measurements, almost no alignment of graphite particles in tetradecane can be observed for DC biased crystals; **b** electrophoretic trapping of Ag-coated PS beads on BSO, distinguished from DEP by the fact that there is no frequency doubling of particles patterns and that alignment occurs in large areas instead of narrow gradient regions (also published in [8])

results applies for sinusoidal and binary amplitude patterns, however, in the case of binary amplitude pattern, longer pearl-chain formation could be observed. This pearl-chain formation is due to the interaction of induced dipoles in an external field and has been reported in many DEP experiments [9, 10]. No influence of the external field frequency could be observed in the range that was tested $100 \leq f \leq 500\,\text{Hz}$, corroborating the finding that for frequencies above the threshold a plateau of constant internal field modulation is reached. In addition to the dielectrophoretic manipulation of neutral particles, electrostatic attraction of charged matter has also been observed for silver-coated polystyrene (PS) micro spheres with a diameter of $2\,\mu\text{m}$. Figure 5.2b shows that the PS beads are attracted to the dark regions of a square pattern incident in the AC-biased crystal. Assuming that electrons are the main charge carriers in BSO, the dark regions can be associated with negative charge, hence the particles must be positively charged [11]. The qualitative behavior in Fig. 5.2b is fundamentally different from what is expected for dielectrophoresis: while DEP occurs at in the presence of high field gradients, hence locally concentrated, electrophoretic attraction is proportional to the charge density and can be distributed over an extended area.

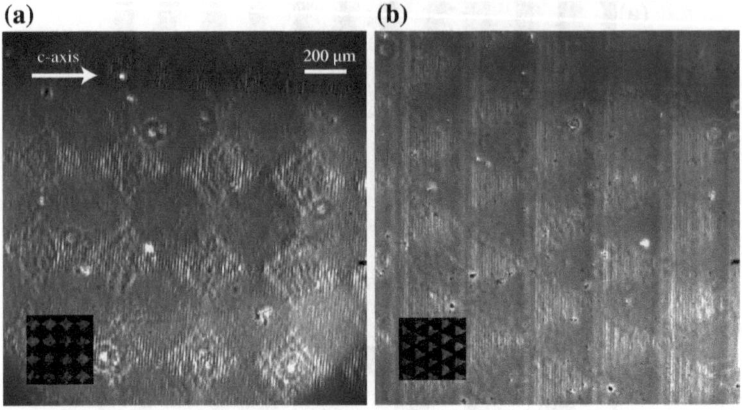

Fig. 5.3 Zernike phase contrast images of refractive index modulation on lithium niobate sample *AT2*; Zernike mode is the appropriate method for qualitative inspection before trapping particles; *insets* show the incident amplitude pattern (also published in [1]) **a** square pattern, **b** triangular pattern

5.2 Lithium Niobate ($LiNbO_3$)

For dielectrophoretic trapping on lithium niobate, less parameters have to be taken into account, which greatly facilitates their use, although the material's conductivity is orders of magnitude smaller. This implies, as measured previously, that even for strong lasers the timescale to induce internal fields is on the order of several seconds to minutes. On the plus side, the material does not require an external voltage supply and—thanks to the low dark conductivity—the internal fields have been measured to be almost constant within a week for crystals with a comparable iron concentration to the ones used in this thesis [12]. This means that the crystal can be moved away from the setup and particles can be blown onto the substrate by an air stream, a method of particle deposition that would not be possible in an optical setup. Figure 5.3 shows the lithium niobate crystal *AT2* in Zernike phase contrast imaging after being structured with binary triangular and square patterns in an external setup at an intensity of $40\,\mathrm{mW\,cm^{-2}}$ for 30 min, so that the crystal has definitely reached its saturation field. Zernike phase contrast is an easy way to inspect the quality of the induced pattern before actually blowing particles over the crystal surface. The images for both trapping patterns show the induced triangular or square structure [1] and can be used to relate the internal pattern with the trapping pattern. The graphite particles ($\phi 6 - 20\,\mu\mathrm{m}$) blown over the surface are clearly trapped at the borders of the dark-white interchanges in the original patterns (cf. Fig. 5.4). Since the crystals had to be removed from the setup, a direct comparison between trapping pattern and particle positions is only possible with the Zernike phase contrast, where dark and bright regions are clearly visible. It is noteworthy that no particles are trapped in the corners of the patterns. A possible reason for this behavior lies in the fact that the

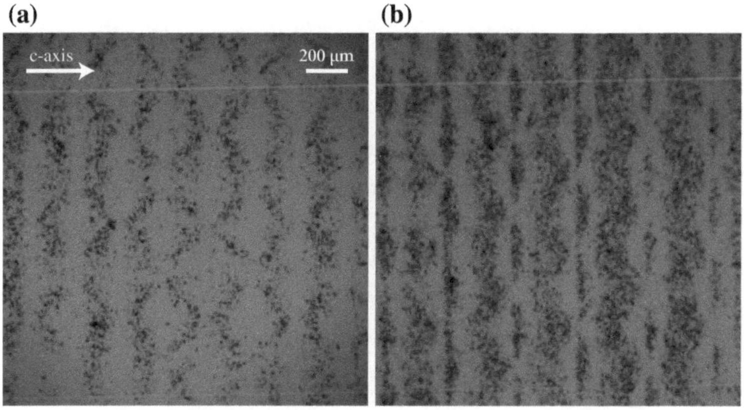

Fig. 5.4 Dielectrophoretic trapping of graphite particles in air on sample *AT2*; particles are trapped at the borders of *dark/white* interchanges in the illumination pattern (also published in [1]) **a** square pattern, **b** triangular pattern

electrode distance in these regions decrease rapidly, implying that the approximation of Eq. 2.3, which only applies for virtual electrode separations much larger than the particle diameter, is not valid any more, so that the assumption of a simple attraction to regions of high field intensity is not valid any more. Additionally, for smaller virtual electrode separations, dipole-dipole interactions have been demonstrated to greatly diminish the trapping efficiency [13].

Fig. 5.5 shows the result of trapping graphite particles in the arbitrary pattern represented by the letters `AG DENZ`. As expected from the material properties of $LiNbO_3$ crystals described in Sect. 3.4, a light modulation perpendicular to the c-axis does not produce sufficient fields, hence field gradients for particle trapping [14], because of the negligible photovoltaic effect along this direction. The pattern of trapped graphite allows an estimation of the maximum inclination angle between pattern k-vector and c-axis. Simply by looking at the letters in Fig. 3.4, one can state that an angle of $\alpha \leq 60°$ must be maintained for a high trapping efficiency. The implications of the anisotropy are further discussed in Sect. 5.3. So far, the writing times for internal fields have been very long. Partially, this can be attributed to the fact that the beam was expanded to illuminate the whole crystal, which might not always be necessary. Furthermore, reduced lithium niobate crystal have been investigated in Sect. 4.1.3, where an increase of the photovoltaic current density of more than one order of magnitude could be measured as compared to as-grown samples if the reduction degree is increased by a thermal treatment. To verify the dielectrophoretic trapping abilities of these reduced crystals, the ASLM was demagnified and coupled into a Nikon TE2000U microscope. For the new demagnification, the total modulator area in the sample plane is only $2.0 \times 1.1\,mm^2$, so that very high mean intensities on the sample surface can be reached. The reduced crystals were illuminated with a binary stripe pattern of period $\Lambda = 422\,\mu m$. Due to the different

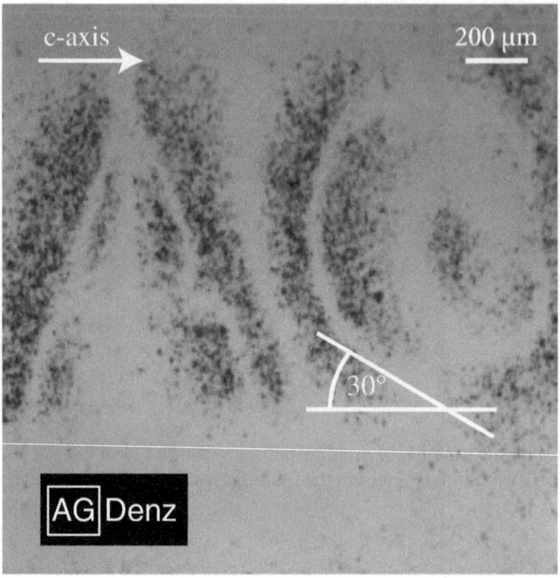

Fig. 5.5 Dielectrophoretic trapping on $LiNbO_3$ in arbitrary patterns; for sufficient DEP forces, the angle between c-axis and the k-vector of the modulation must not exceed 60° (also published in [1])

thicknesses and absorption coefficients, special care was taken that the top surface intensity of 233 mW cm^{-2} is the same for all crystals, since the electric fields near the surface are mainly responsible for the trapping [15, 16]. The illumination time has been reduced according to the increase in the photovoltaic current density that has been measured in Fig. 4.2b: from 30 s for the as-grown sample *AT1*, to 9 s for sample 151.2 and 3 s for samples 151.6.1 and 151.5.1. The illumination time for the last samples was chosen to be the same since the photovoltaic current density saturates for samples above a reduction degree of 1, which is the case for 151.5.1 with a reduction degree of 1.63 [3]. The result depicted in Fig. 5.6 shows that for the first three samples, the trapping efficiency is constantly high despite the fact that the writing time has been reduced significantly. For the sample with a reduction degree of 1.63 in Fig. 5.6d, the overall trapping efficiency is significantly reduced and the trapping contrast, i.e. the ratio between particles in desired and undesired locations is lower as compared to the other samples. This can be explained by the reduced internal saturation field of the highly reduced samples which is more than one order of magnitude lower and the stagnating photovoltaic current density. Summing up the findings from these trapping experiments, it can be concluded that samples up to a reduction degree of 0.74 can be efficiently used for POT. For the more reduced samples, which would theoretically be expected to have an even larger photoconductivity, the higher conductivity is accompanied by a disproportional decrease in the electric field strength, which prohibits their use for POT.

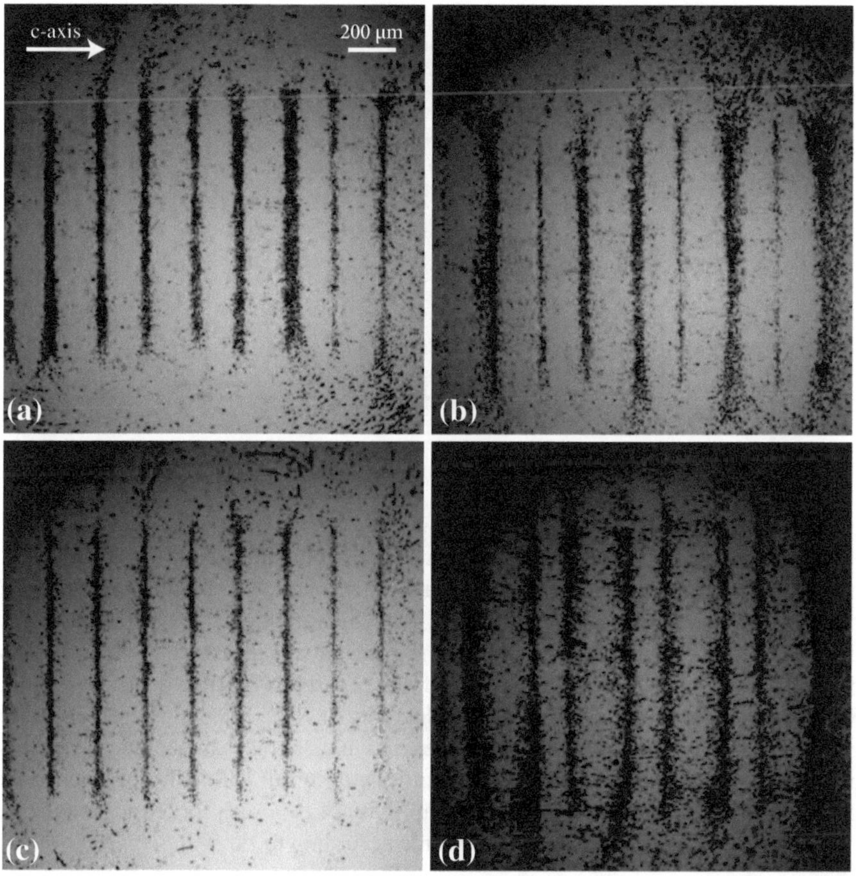

Fig. 5.6 Reduced $LiNbO_3$ samples for POT; all samples were illuminated with a top surface intensity of 233 mW cm^{-2}; the illumination time was decreased as denoted in the subcaptions; for the most reduced sample 151.5.1 with $r = 1.63$, a significant decrease in the trapping efficiency could be observed (also published in [3]) **a** *AT1*, $t_{\text{illum}} = 30$ s, **b** *151.2*, $t_{\text{illum}} = 9$ s, **c** *151.6.1*, $t_{\text{illum}} = 3$ s, **d** *151.5.1*, $t_{\text{illum}} = 3$ s

5.3 Measurement and Anisotropy of Dielectrophoretic Forces in POT

In Chap. 2, it was shown that in the case of a dipole approximation for spherical particles, the dielectrophoretic force is directly proportional to the gradient of the electric field intensity, a value that can be quantified using DHM. A typical example for such a measurement can be found in Fig. 5.7. In this measurement, the $LiNbO_3$ crystal *AT1* was illuminated with a binary stripe pattern and the internal field intensity gradient g was calculated as $g(k) = (E^2(k) - E^2(k+1))/\Delta x$, where Δx is the separation between pixels in the object domain and k is the pixel index. In general,

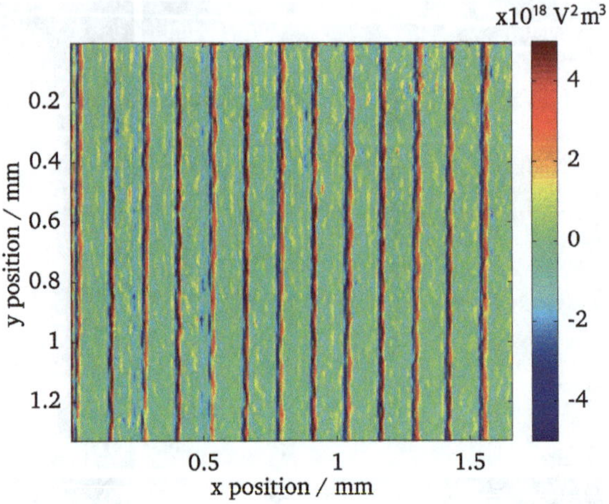

Fig. 5.7 Direct measurement of internal field gradients in $LiNbO_3$ for binary stripe patterns; from the calculation of the field intensity gradient in the saturation case, the DEP force near the crystal surface can be inferred

it can be concluded that very high gradients occur at the interchanges between dark and bright regions in the illumination pattern. In particular, for the binary stripe pattern with its k-vector along the c-axis, for which the highest possible gradients can be expected, the maximum value of the gradient was measured to be as high as $5\times10^{18}\,V^2\,m^{-3}$. For the exemplary case of glassy carbon (GC) powder ($\phi 2-12\,\mu m$, Sigma-Aldrich), which has a specific conductivity of $700\,S\,cm^{-1}$ [17], hence several orders of magnitude larger than for alkanes, to which tetradecane belongs [18], this allows to approximate the Clausius-Mossotti factor in Eq. 2.13 as 1. For a spherical particle of $r = 1\,\mu m$ and the relative permittivity of $\varepsilon_r = 2\varepsilon_0$ for tetradecane [19], this allows the direct calculation of the force according to Eq. 2.9:

$$F_{\text{DEP}} = 4\pi(1\times10^{-6}\,\text{m})^3\varepsilon_0 5\times10^{18}\,\text{V}^2\,\text{m}^{-3}$$
$$= 0.556\times10^{-9}\,\text{N}. \quad (5.1)$$

Note that this force is already larger than that of the strongest OT [20] and increases with the cube of the particle size. It has to be mentioned that this calculation of the force is only valid in the immediate vicinity to the surface. Burgos et al. have calculated the force components perpendicular and parallel to the crystal surface separately and state that the dielectrophoretic force decays in the direction normal to the surface with $e^{-\Lambda d}$ where Λ is the pattern period and d is the distance from the surface [16]. Furthermore it must be taken into account that the presented measurement of the field gradients represent an average along the thickness of the crystal. It is therefore possible that the gradient near the surface is even higher than the one calculated in this section, due to the fact that the pattern cannot be imaged sharply through the whole crystal. A possible solution for this predicament is the use of a diffusion-doped $LiNbO_3$ sample, which is further described in the outlook.

So far, no attention has been paid to the acquisition of phase gradients in different geometries, where the k-vector of the modulation is not along the polar axis of the material. Figure 5.5 showed that the dielectrophoretic trapping is significantly reduced or even completely prohibited if the angle between the k-vector of the intensity modulation and the c-axis of lithium niobate is larger than 60°, a value that could be inferred from the tangential line to the rounded shape of the letter G (cf. Fig. 5.5). The reason for this behavior lies in the anisotropy of the charge transport of the photovoltaic effect as well as drift. It has been described in Sect. 3.4 that the main contribution to the charge transport phenomena is along the c-axis whereas the movement of charge carriers perpendicular to this distinguished axis can be neglected. This results in very high electric field gradients for any modulation along the c-axis but not perpendicular to it. In order to verify this finding, the previously described setup for structuring the photorefractive crystal in-situ, i.e. in the microscope sample plane, is extended by a DHM setup which was integrated into the Nikon TE2000U microscope. This combination allows to directly measure the electric field gradients and the resulting confinement of graphite test particles. The crystal 151.6.1 was illuminated with a binary amplitude grating, whose orientation was rotated from 0° to 80° in steps of 10°. The illumination time for all gratings was 30 s at an intensity of 1.2 W cm^{-2} incident on the bottom of the crystal. After the digital holographic measurements of the internal fields, the test solution of graphite particles in tetradecane was dripped onto the sample. Figure 5.9 shows the exemplary results from three data points in order to corroborate the findings from the previous experiments. It can be nicely seen that for a grating angle of 40°, the relatively high field gradients are able to trap the graphite particles in a precise fashion replicating the original structure. For $\alpha = 60°$, however, the electric field gradients are greatly reduced and hence are barely sufficient to capture particles in the predefined pattern. This case corroborates the previous finding of the maximum angle between c-axis and k-vector. For an angle of 70°, it can be concluded that the trapping powers are insufficient, which can be seen in the visualization of ∇E^2 as well as in the resulting particle trapping image. It is noteworthy that the results from Fig. 5.5 have been derived on an as-grown crystal, whereas the test in this section was arranged on a reduced crystal with a reduction degree of $r = 0.36$. The result that the minimum angle is independent of the crystal composition might appear surprising at first because the strength of the trapping power should be connected to the maximum field that can be achieved and which is expected to be highly-dependent on the reduction degree. However, the findings become clearer when taking into account the previous results that both the as-grown commercial crystal and sample 151.6.1 have comparably high saturation fields. It can be concluded that a maximum angle of 60° applies for the crystals up to a reduction degree of $r = 0.36$ whereas it must be even smaller for crystal with a diminished field. Interestingly, even for BSO, where the internal field is approximately one order of magnitude lower than for $LiNbO_3$, trapping could be observed for binary illumination patterns that were rotated up to 45°. Figure 5.8 shows that particles align in the desired rotated geometry, while the pearl chain formation is governed by the electric field direction, which is from left to right. This can be easily explained by the fact that pearl-chain formation takes place already in the bulk fluid where the particles

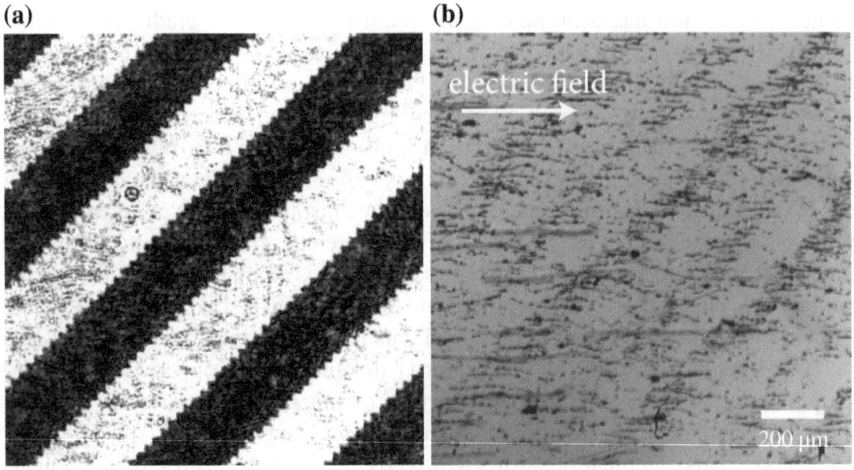

Fig. 5.8 Dielectrophoretic trapping of graphite flakes in tetradecane for rotated patterns on BSO; although the electric field is much lower than for $LiNbO_3$, trapping could be observed up to an angle of 45° between trapping pattern and electric field direction **a** incident pattern, **b** trapped graphite particles

do not sense the inhomogeneous electric field of the BSO, but are only polarized by the external field.

5.4 Surface Discharge Model

A notable thing about the[1] trapping experiments that have been published on photorefractive substrates so far is that they were all achieved with either air or oily substances as the surrounding medium. Many of the publications used tetradecane [1, 2], other authors have successfully implemented pentanoic acid [21], paraffin oil [14] and silicone oil [22] as the suspension medium.
From the theoretical consideration of the dielectrophoretic force, it has been shown in Eq. 2.13 that attractive forces will occur whenever the particle under investigation is more conductive than the surrounding medium. For glassy carbon spheres, this condition is fulfilled for almost any substance, including water. In order to derive which solvents can be used especially in $LiNbO_3$-based OET and how the solvent conductivity changes the DEP behavior, the charge transport processes near the crystal surface in the presence of a conducting medium can be described with the help of the equivalent circuit sketched in Fig. 5.10. The charge separation reached through the bulk photovoltaic effect (or drift in an external field) can be approximated by the charging process of a capacitor C over the resistance R_1, which is the resistivity

[1] Results partly obtained in collaboration with Stefan Gläsener during his Diploma thesis.

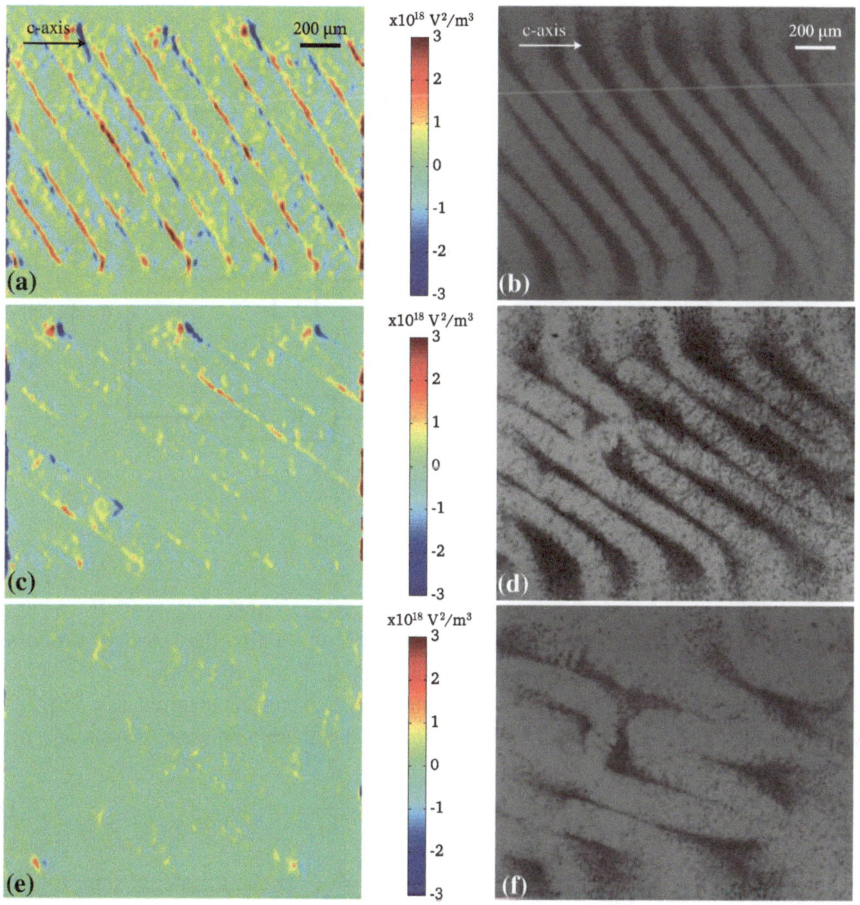

Fig. 5.9 Field intensity gradient and trapped particles for a binary pattern with different angles to the c-axis of the lithium niobate crystal 151.6.1; only for modulations with an angle between k-vector and c-axis of $\leq 60°$, satisfactory trapping can be observed; test solution consists of glassy carbon particles in tetradecane **a** ∇E^2 for 40°, **b** trapped GC particles for 40°, **c** ∇E^2 for 60°, **d** trapped GC particles for 60°, **e** ∇E^2 for 70°, **f** trapped GC particles for 70°

of the crystal and its discharge over a resistor R_2, which corresponds to the resistivity of the liquid medium. According to Kirchhoff's circuit laws, the currents must fulfill the condition

$$I_{R_1} - I_{R_2} - I_C = 0. \tag{5.2}$$

Kirchhoff's mesh rule further yields that for the described system:

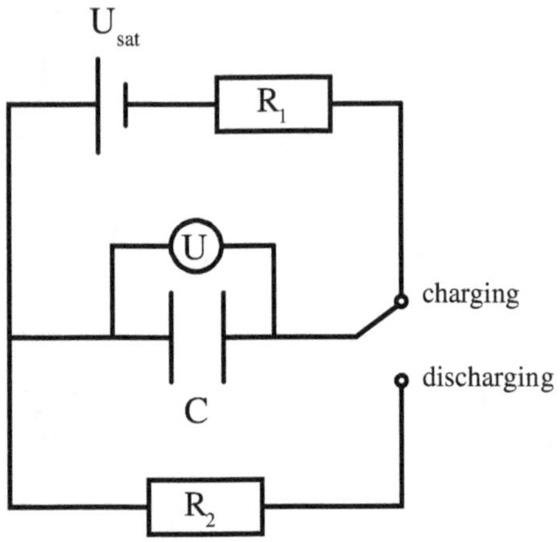

Fig. 5.10 Equivalent circuit for the charge separation on a photorefractive surface in the presence of a conducting medium

$$U_{\text{sat}} = U_{R_1} + U_C \quad \text{and}$$
$$U_{R_2} = U_C \tag{5.3}$$

where U_{sat} is the saturation voltage of the bulk photovoltaic effect or drift process, which serves as the "voltage supply" in this model. Using the relation $I_C = C \times dU_C/dx$ for the current through a capacitor, Eq. 5.3 can be inserted into Eq. 5.2, which gives a linear differential equation for the capacitor voltage:

$$\frac{dU_C(t)}{dt} = \frac{U_{\text{sat}}}{\tau_{\text{ch}}} - \frac{U_C(t)}{\tau_0} \tag{5.4}$$

where the new variables $\tau_{\text{ch}} = R_1 C$ and $\tau_{\text{dis}} = R_2 C$ as the time constants for the charging and discharging processes and the abbreviation $\tau_0 = \tau_{\text{ch}}\tau_{\text{dis}}/(\tau_{\text{ch}} + \tau_{\text{dis}})$ have been introduced. This inhomogeneous differential equation is solved by the homogeneous solution and the method of variation of parameters, which leads to the general solution for $U_C(t)$:

$$U_C(t) = \frac{U_{\text{sat}}\tau_{\text{dis}}}{\tau_{\text{ch}} + \tau_{\text{dis}}} + de^{-\frac{t}{\tau_0}} \tag{5.5}$$

where the constant d is determined by the boundary conditions. The first case to be considered is that of a "fully charged" capacitor ($U_C(0) = U_{\text{sat}}$), i.e. a crystal that has already reached its saturation field. The final solution is

$$U_C(t) = \frac{U_{\text{sat}}\tau_{\text{dis}}}{\tau_{\text{ch}} + \tau_{\text{dis}}} + \frac{U_{\text{sat}}\tau_{\text{ch}}}{\tau_{\text{ch}} + \tau_{\text{dis}}}\mathrm{e}^{-\frac{t}{\tau_0}} \tag{5.6}$$

This solution allows the differentiation between two limiting cases: if the conductivity of the liquid medium is much larger than that of the crystal, then $\tau_{ch} \gg \tau_{dis}$ and $\tau_0 \approx \tau_{dis}$, so that Eq. 5.6 can be simplified to

$$U_C(t) \approx U_{sat} e^{-\frac{t}{\tau_{dis}}} \tag{5.7}$$

which describes a fast decay of the induced space-charge field. On the other hand if the conductivity of the crystal is higher than that of the medium, $\tau_{dis} \gg \tau_{ch}$ and $\tau_0 \approx \tau_{ch}$, in which case Eq. 5.6 reduces to

$$U_C(t) \approx U_{sat} \tag{5.8}$$

which means that the internal field stays essentially constant. A different set of boundary conditions arises, if the crystal is uncharged, but already covered with liquid when the light is turned on, i.e. the charging process starts. In this case $U_C(0) = 0$ and the total solution is

$$U_C(t) = \frac{U_{sat}\tau_{dis}}{\tau_{ch} + \tau_{dis}} \left\{1 - e^{-\frac{t}{\tau_0}}\right\} \tag{5.9}$$

In this case, the crystal is charged similarly to the previous description, but only to a fraction of its original saturation value, as determined by the ratio $\tau_{dis}/(\tau_{ch} + \tau_{dis})$. This results serves as an explanation why all experiments so far have presented dielectrophoretic trapping in the presence of insulating media. For example, if water would be used, even the best distilled water still has a very high conductivity of $5 \times 10^{-8}\,\mathrm{S\,cm^{-1}}$ [23], as compared to a typical value of the specific photoconductivity $\sigma_{ph}/(Ir) = 1 \times 10^{-12}\,\mathrm{S\,cm\,W^{-1}}$ for iron-doped samples [24, 25], where I is the incident intensity and r is the reduction degree of the material, hence the ratio $c_{Fe^{2+}}/c_{Fe^{3+}}$. Neglecting dark conductivity, it can be concluded that even for comparably high light powers of $1\,\mathrm{W\,cm^{-2}}$ and a reduction degree of unity, the conductivity for example of lithium niobate is in the range of $10^{-12}\,\mathrm{S\,cm^{-1}}$, which is four orders of magnitude lower than that of water. A similar calculation can be made for the case of BSO. Although the photoconductivity here is three orders of magnitude higher, a significantly higher conductivity than the one experienced in water cannot be reached. In any case, the field would be greatly reduced as described in Eq. 5.9 (Table 5.1).

Since the discharge process is only limited to a thin layer near the surface of the crystal, an influence on the phase retardation of the crystal could not be seen when discharged crystals were probed by DHM. Nevertheless, the previous assumptions can be tested using solvents of different conductivity. As mentioned above, the use of water is prohibited by its comparably high conductivity. Yet, other substances whose resistivity is far less or comparable to that of $LiNbO_3$ may be used. Figure 5.11 shows a series of experiments performed by Stefan Gläsener during his diploma thesis [26]). The $LiNbO_3$ crystal *AT2* was structured for 4 min with a sinusoidal pattern of $I = 160\,\mathrm{mW\,cm^{-2}}$. Afterwards, different solvents of decreasing conductivity

Table 5.1 Conductivities of solvents used in this thesis

Liquid	$\sigma/\mathrm{S\,cm^{-1}}$
Acetone	5.98×10^{-7} [28]
Iso-2-propanole	5.7×10^{-8} [29]
Water	5×10^{-8} [23]
Toluene	3.4×10^{-12} [28]
n-Hexane	$<1 \times 10^{-14}$ [30]
Tetradecane	$<1 \times 10^{-15}$

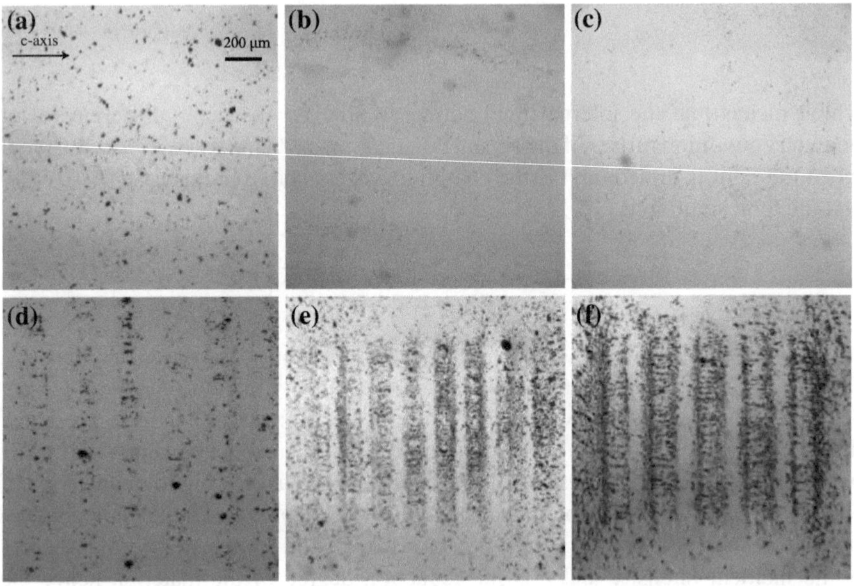

Fig. 5.11 Dielectrophoretic particle experiments with solvents of decreasing conductivity; for solvent conductivities significantly higher than that of the crystal, as in Fig. 5.11a–c, no DEP can be observed due to a discharge of surface fields; for media with lower conductivity, particle alignment can clearly be observed (Images obtained by Stefan Gläsener [26]) **a** acetone, **b** water, **c** isopropanole, **d** toluene, **e** n-hexane, **f** tetradecane

with suspended homemade graphite flakes were dripped onto the surface. For all solvents, it is a safe assumption that the graphite has a significantly higher conductivity than the liquid medium [27], so attractive dielectrophoretic forces are expected in all cases. The result (Fig. 5.11a–c) clearly shows that for the media with a relatively high conductivity, no forces are observed at all and the graphite particles sediment randomly, which is a strong indication for the hypothesis that the space-charge fields near the crystal surface have been fully discharged before the particles actually reach the crystal surface. On the other hand, for the media with a conductivity comparable to or even less than the $LiNbO_3$ substrate, attractive forces can be observed, the clarity of the pattern increasing with increasing medium resistivity.

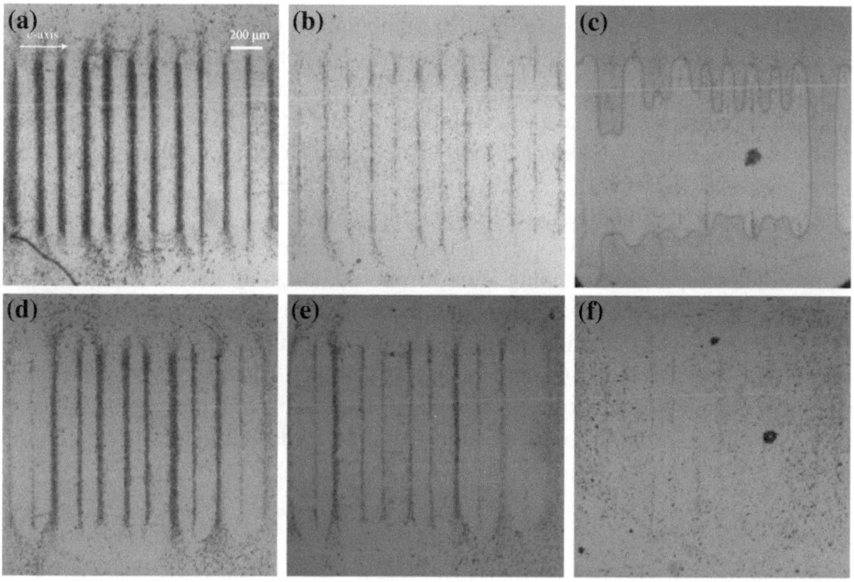

Fig. 5.12 Influence of crystal surface discharge on DEP trapping efficiency for n-hexane and tetradecane (TD), respectively; subcaptions indicate the pause between the light-induced structuring of crystal 151.6.1 and the adding of particles; for more details about the measurement protocol refer to the text **a** n-hexane: 0 s, **b** after 90 s, **c** after 120 s, **d** TD: 0 s, **e** after 30 min, **f** after 60 min

If this result is really caused by the medium conductivity and not by any other property, the model developed earlier suggests that a temporal decrease of the trapping efficiency should be visible in the experiments. Additionally, according to Eq. 5.9, it should be possible to induce charges on the surface of the crystal even in the presence of a medium. For Fig. 5.12a–c, the crystal 151.6.1 has been covered with pure hexane and illuminated with a binary stripe pattern of $I = 0.89\,\mathrm{W\,cm^{-2}}$ for 10 s, which corresponds to the charging process. The presence of the pure solvent ensures that no particles are trapped while the crystal surface is constantly discharged. After a certain discharge time t_{dis}, GC particles suspended in the same liquid are added to the crystal surface. The experiment is performed several times, with an increasing interval between optical structuring and adding dispersed particles, as indicated by the subcaptions. For comparison, the very same experiment was performed for a crystal which was covered with tetradecane (cf. Fig. 5.12d–f). Although the total density of trapped particles varies slightly from experiment to experiment, the assumption of a decreasing trapping efficiency is clearly corroborated by the findings of Fig. 5.12. Since liquid carbohydrates, such as hexane and tetradecane, are themselves not electrically conducting, their conductivity is supposed to be based solely on the presence of trace impurities [18]. Therefore, the real conductivities are too low to be measured with off-the-shelf equipment. An impedance spectroscopy measurement in the group of Prof. Schönhoff in the *Institute of Physical Chemistry* yielded an upper limit of $\sigma_{\mathrm{tetradecane}} < 1 \times 10^{-15}\,\mathrm{S\,cm^{-1}}$, which was the limit of accuracy of the measurement device. For highly-purified hexane, other authors have measured an upper limit of $\sigma_{\text{hexane}} < 1 \times 10^{-14}\,\text{S}\,\text{cm}^{-1}$ [30]. Based in the assumption that this conductivity is really based on trace impurities, a higher conductivity for hexane would be expected since the impurities in hexane have a significantly higher mobility due to the low viscosity. In comparison, the longer carbohydrate chains in tetradecane are responsible for an increased viscosity [31]. Consequently, for hexane, which is supposed to have a significantly higher conductivity than tetradecane, the trapping is greatly reduced already after 90 s of liquid discharging the crystal surface, whereas tetradecane still permits dielectrophoretic trapping after 30 min. Although all these findings as well as the references cited above corroborate the model presented for the discharge of surface charges, a critical view on the medium conductivities is necessary. All the media were dispensed with standard laboratory syringes, which were used only once, but otherwise not specially treated. It is therefore possible that the conductivities of the solvent used in the experiment differ from the measured values in high-purity samples.

References

1. M. Esseling, F. Holtmann, M. Woerdemann, C. Denz, Two-dimensional dielectrophoretic particle trapping in a hybrid crystal/PDMS-system. Opt. Express **18**(16), 17404–17411 (2010)
2. S. Glaesener, M. Esseling, C. Denz, Multiplexing and switching of virtual electrodes in opto-electronic tweezers based on lithium niobate. Opt. Lett. **37**(18), 3744–3746 (2012)
3. M. Esseling, A. Zaltron, N. Argiolas, G. Nava et al., Highly reduced iron-doped lithium niobate for opto-electronic tweezers. Appl. Phys. B Lasers Opt. **113**(2), 191–197 (2013)
4. M. Esseling, A. Zaltron, C. Sada, C. Denz, Charge sensor and particle trap based on z-cut lithium niobate. Appl. Phys. Lett. **103**, 061115 (2013)
5. M. Woerdemann, C. Alpmann, M. Esseling, C. Denz, Advanced optical trapping by complex beam shaping. Laser Photonics Rev. **7**(6), 839–854 (2013)
6. K. Kang, X. Xuan, Y. Kang, D. Li, Effects of DC-dielectrophoretic force on particle trajectories in microchannels. J. Appl. Phys. **99**(6) (2006)
7. Y. Kang, D. Li, S.A. Kalams, J.E. Eid, DC-dielectrophoretic separation of biological cells by size. Biomed. Microdevices **10**(2), 243–249 (2008)
8. M. Esseling, S. Glaesener, F. Volonteri, C. Denz, Opto-electric particle manipulation on a bismuth silicon oxide crystal. Appl. Phys. Lett. **100**(16), 161903 (2012)
9. K. Hermanson, S. Lumsdon, J. Williams, E. Kaler et al., Dielectrophoretic assembly of electrically functional microwires from nanoparticle suspensions. Science **294**(5544), 1082–1086 (2001)
10. R. Kretschmer, W. Fritzsche, Pearl chain formation of nanoparticles in microelectrode gaps by dielectrophoresis. Langmuir **20**(26), 11797–11801 (2004)
11. M. Petrov, V. Bryksin, in *Photorefractive Materials and Their Applications 2: Materials*, ed. by P. Günter, J.-P. Huignard. Space-Charge Waves in Sillenites: Rectification and Second-Harmonic Generation (Springer, New York, 2007), pp. 285–325
12. I. Nee, M. Muller, K. Buse, E. Kratzig, Role of iron in lithium-niobate crystals for the dark-storage time of holograms. J. Appl. Phys. **88**(7), 4282–4286 (2000)
13. N. Markarian, M. Yeksel, B. Khusid, K. Farmer et al., Limitations on the scale of an electrode array for trapping particles in microfluidics by positive dielectrophoresis. Appl. Phys. Lett. **82**(26), 4839–4841 (2003)

14. H.A. Eggert, F.Y. Kuhnert, K. Buse, J.R. Adleman et al., Trapping of dielectric particles with light-induced space-charge fields. Appl. Phys. Lett. **90**, 241909 (2007)
15. J. Villarroel, H. Burgos, A. Garcia-Cabanes, M. Carrascosa et al., Photovoltaic versus optical tweezers. Opt. Express **19**(24), 24320–24330 (2011)
16. H. Burgos, M. Jubera, J. Villarroel, A. Garcia-Cabanes et al., Role of particle anisotropy and deposition method on the patterning of nano-objects by the photovoltaic effect in LiNbO3. Opt. Mater. **35**(9), 1700–1705 (2013)
17. Sigma-Aldrich. Carbon glassy, spherical powder, 2–12 μm, 99.95 metals basis (2013), http://www.sigmaaldrich.com/catalog/product/aldrich/484164?lang=en&region=
18. D. Swan, Review of recent investigations into electrical conduction and breakdown of dielectric liquids. Br. J. Appl. Phys. **13**, 208 (1962)
19. C. Rosenberg, N. Hermiz, R. Cook, Cavity resonator measurements of the complex permittivity of low-loss liquids. IEEE Proc. H Microwaves Antennas Propag. **129**(2), 71–76 (1982)
20. N. Malagnino, G. Pesce, A. Sasso, E. Arimondo, Measurements of trapping efficiency and stiffness in optical tweezers. Opt. Commun. **214**(1–6), 15–24 (2002)
21. S. Grilli, P. Ferraro, Dielectrophoretic trapping of suspended particles by selective pyroelectric effect in lithium niobate crystals. Appl. Phys. Lett. **92**, 232902 (2008)
22. X. Zhang, J. Wang, B. Tang, X. Tan et al., Optical trapping and manipulation of metallic micro/nanoparticles via photorefractive crystals. Opt. Express **17**, 9981–9988 (2009)
23. Millipore Water Purification Systems. Milli Q reference ultrapure water purification system (2013), http://www.millipore.com/catalogue/module/c79625
24. K. Peithmann, J. Hukriede, K. Buse, E. Kratzig, Photorefractive properties of LiNbO3 crystals doped by copper diffusion. Phys. Rev. B Condens. Matter Mater. Phys. **61**(7), 4615–4620 (2000)
25. K. Buse, J. Imbrock, E. Krätzig, K. Peithmann, in *Photorefractive Materials and their Application*, ed. by P. Günter, J.-P. Huignard. Photorefractive Effects in LiNbO3 and LiTaO3 (Springer, New York, 2007), pp. 83–126
26. S. Glaesener. Mikromanipulation mit einer $LiNbO_3$-basierten optoelektrischen Pinzette, Diploma thesis, University of Münster (2012)
27. N. Deprez, D. McLachlan, The analysis of the electrical-conductivity of graphite powders during compaction. J. Phys. D Appl. Phys. **21**(1), 101–107 (1988)
28. L. Liu, Y. Yang, Y. Zhang, A study on the electrical conductivity of multi-walled carbon nanotube aqueous solution. Physica E Low dimens. Syst. Nanostruct. **24**(3–4), 343–348 (2004)
29. M. Prego, E. Rilo, E. Carballo, C. Franjo et al., Electrical conductivity data of alkanols from 273 to 333 K. J. Mol. Liq. **102**(1–3), 83–91 (2003)
30. E. Forster, Electric conduction in liquid hydrocarbons: 3. Comparison of saturated + unsaturated hydrocarbons. J. Chem. Phys. **40**(1), 91 (1964)
31. J. Dymond, H. Oye, Viscosity of selected liquid n-alkanes. J. Phys. Chem. Ref. Data **23**, 41–53 (1994)

Chapter 6
Optofluidic Applications for Photorefractive Optoelectronic Tweezers

In this chapter, advanced applications of the previously described POT will be presented. These applications are especially relevant for the field of optofluidics, which describes the fusion of optical elements with the beneficial aspects of microfluidics [1–3]. Two of the demonstrator concepts will exploit the specific properties of $LiNbO_3$ crystal, once by using the long dark storage time of inscribed patterns, and by employing the well-characterized unidirectional charge transport along the c-axis. This property can be used to demonstrate arbitrary electrophoretic trapping on a z-cut surface without any restrictions due to crystal anisotropy. The last two concepts make use of the fact, that apart from the ability to handle myriads of particles, the occurrence of dielectrophoretic forces is not limited to discrete particles but is as well able to manipulate continuous media like polymers or fluids. This behavior will be used to pattern polymeric material prior to the curing process in order to fabricate soft diffraction gratings. In the last application, a $LiNbO_3$ niobate substrate will be used to construct an optofluidic router, which is able to steer, hold and sort microfluidic droplets by its optically-induced electric field gradients. Please note that while the first two concepts are limited to $LiNbO_3$ and materials with similar photoconductive and photovoltaic properties like, for example, lithium tantalate [4], the last two concepts can in in principle be realized on any photoconductive surface. The reason why $LiNbO_3$ is also used here is due to its outstanding ability to generate high internal space-charge fields without the need for an external voltage supply. It goes without saying that a material with these properties incredibly simplifies microscopic integration.

For three out of the four concepts that will be presented in this last chapter, polydimethylsiloxane (PDMS) micro structures will play an important role, so the interested reader is referred to Appendix B, where the basic properties of PDMS are summarized and the protocol to fabricate custom-made microchannel masters is described. Nevertheless, comprehensive knowledge about PDMS is not necessarily required for this chapter, other than the fact that PDMS is a polymer that can be conveniently cast into any shape in its liquid state before curing. After curing, it forms a visco-elastic solid that can be peeled off the master structure and is

M. Esseling, *Photorefractive Optoelectronic Tweezers and Their Applications*, Springer Theses, DOI 10.1007/978-3-319-09318-5_6 self-adherent to a variety of surfaces. Due to its good optical and mechanical properties, it has been the material of choice for many microfluidic applications.

6.1 Multiplexing and Switching of Virtual Electrodes

Apart from its high internal saturation field and the ensuing refractive index change, the long dark storage time has been one of the main appeals to use $LiNbO_3$ samples for the realization of holographic data storage systems. Due to the low dark conductivity, it takes days to months for the internal field to decay. Consequently, virtual electrodes that have been inscribed into a POT substrate will remain even after the illumination is switched off, provided that they are not erased by strong ambient illumination or elevated temperature. In this section, as-grown $LiNbO_3$ crystals will be used for the sequential induction and erasure of internal fields. The long storage time is exploited to create complex electrode configurations with relatively simple light patterns, whose location on the crystal is changed sequentially. This approach is referred to as *spatial multiplexing* [5]. Figure 6.1 shows the results of the sequential large-scale induction of a stripe-like electric field structure and the respective trapping of glassy carbon beads in tetradecane in a three-step procedure. A binary stripe-shaped light field was created with the ASLM and imaged to three different positions on the crystal, which was integrated into the Nikon TE2000U microscope. The mean intensity during writing was $I = 287\,\mathrm{mWcm^{-2}}$. As was demonstrated before in Sect. 5.3, the electric field gradients along the c-axis are much stronger than those perpendicular to it. The writing time for the different electrodes was adjusted according to their angle with the c-axis to partly compensate this; for the 90° electrodes it was set to $t_1 = 60\,\mathrm{s}$ and $t_2 = \sqrt{2}t_1 \approx 85\,\mathrm{s}$ for the 45° electrode.

To observe sequential particle trapping, a PDMS reservoir is fabricated, UV/ozone treated and then applied to the crystal surface and filled with a suspension of tetradecane/glassy carbon. The UV/ozone treatment has been demonstrated to prevent the diffusion of non-polar solvents (such as tetradecane) into the bulk PDMS [6]. Since the first virtual electrode is created during particle sedimentation, the moving GC particles in the fluid volume are influenced by the evolving field gradients and trapped at the electrode edges as described by the theory about dielectrophoresis. For the further electrodes, particle reorganization is inhibited due to friction and adhesion to the crystal surface. Therefore, the fluid flow is increased with a pipette and adhered particles are dispersed again and are attracted by the newly created field gradients. As expected from the anisotropy of the photovoltaic effect, particle trapping at electrode edges not perpendicular to the c-axis is less pronounced despite the longer writing time of 85 s. In the gap of two facing electrode edges, the typical effect of pearl-chain formation can be observed [7]. It results from the fact that neighboring GC particles are polarized collinearly, hence the two oppositely polarized ends of the beads attract each other. This experiment has demonstrated that it is indeed possible to sequentially induce a multitude of virtual electrodes without affecting the previously captured particles. Such a concept has possible applications in fields, where a very

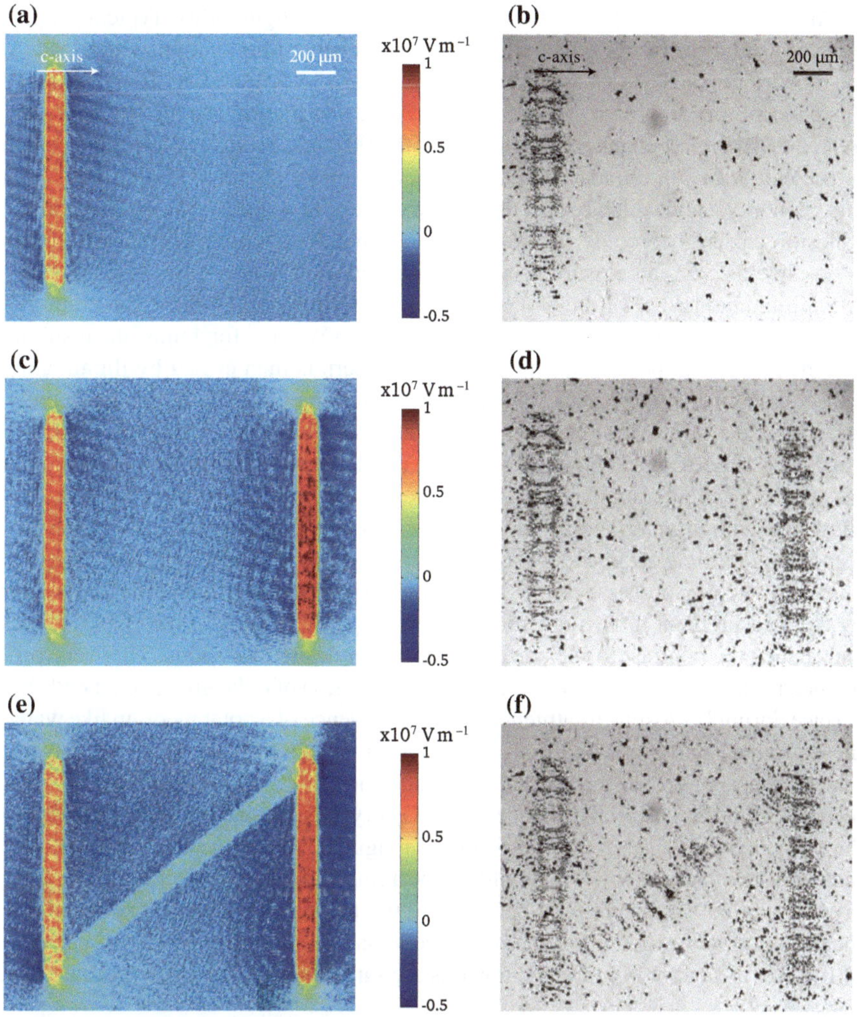

Fig. 6.1 Sequential induction of virtual electrodes in as-grown $LiNbO_3$ crystal and subsequent trapping of glassy carbon beads suspended in tetradecane (also published in [5]). **a** electric field after step 1; **b** particle trapping after step 1; **c** electric field after step 2; **d** particle trapping after step 2; **e** electric field after step 3; **f** particle trapping after step 3

large surface is to be structured with trapped matter [8]. Normally, there is a trade-off between the induction of a high resolution pattern that is created by adequately demagnifying a spatial light modulator, or a large scale induction of patterns where the light modulator is not or only slightly demagnified. The spatial multiplexing approach allows the repeated induction of a multitude of high resolution patterns, thereby offering a way out of the previously described trade-off.

Another beneficial aspect of POT is their optical reconfigurability. Typically, light-induced virtual electrodes can be erased by homogeneous illumination, which leads to the redistribution of charge carriers and the erasure of electric fields. To demonstrate this concept, the as-grown crystal is employed as the bottom sealing for an ozone-treated PDMS microchannel (height/width:$50/300\,\mu m$), through which a suspension of graphite flakes in tetradecane was pumped at a flow rate of $V \approx 10\,\mu L\,s^{-1}$. In this geometry a controllable fluid flow occurs naturally and the previously described adhesion can be overcome. In order to show that switching of virtual electrodes is possible as well, an erasure process was established between two writing steps. After illuminating the bottom of the channel with a three-period sinusoidal grating of $\Lambda = 200\,\mu m$ at an incident mean intensity of $250\,m\,Wcm^{-2}$ for 4 min , the resulting alignment that can be seen in Fig. 6.2b. The pattern is then erased by illuminating the crystal with the microscope's incandescent illumination for 6 min. Subsequently, a second electrode writing process of the same structure, but at an angle of 45° to the c-axis is performed. By now it should be clear that the trapping efficiency in the second configuration is reduced compared to the first one. However, Fig. 6.2c, d show that with an intermediate erasure process electric field gradients and the corresponding particle arrangements can be partially removed and switched. The total number switching processes is not limited by the photorefractive material, but rather by the measurement time, since more and more particles stick to the channel walls and reduce the contrast to the particles trapped in desired locations. This very pronounced behavior may be explained by the introduction of polar groups on the PDMS surface during UV/ozone treatment [9]. In the presence of a polar medium like water, these groups can be screened to the particles, but not in the case of apolar solvents like tetradecane, so that there is increased adhesion to the PDMS. Nevertheless, for increased light intensity and reduced $LiNbO_3$ crystals that have a higher photoconductivity, experiments can be performed in significantly reduced time frames, so that adhesion to channel walls should not be a limiting factor. Especially the erasure process can be significantly be sped up by the use of directed LED sources. The incandescent halogen lamp of the microscope has a total power of 100 W but most of it is lost as heat or not directed towards the samples and never reaches it.

6.2 Charge Sensing and Particle Trapping on z-Cut Lithium Niobate Samples

In all previous experiments, the direction of the distinguished axis of charge transport, that is determined either by the direction of the c-axis ($LiNbO_3$) or the direction of the applied electric field (BSO, drift-dominated materials) has been along the crystal surface. This ensures an efficient charge separation and leads to highly-modulated fields with very sharp field gradients (see also Sect. 5). In all of these applications, a challenge is posed by the anisotropy of the charge transport, which results in a negligible trapping efficiency for grating vectors perpendicular to the distinguished

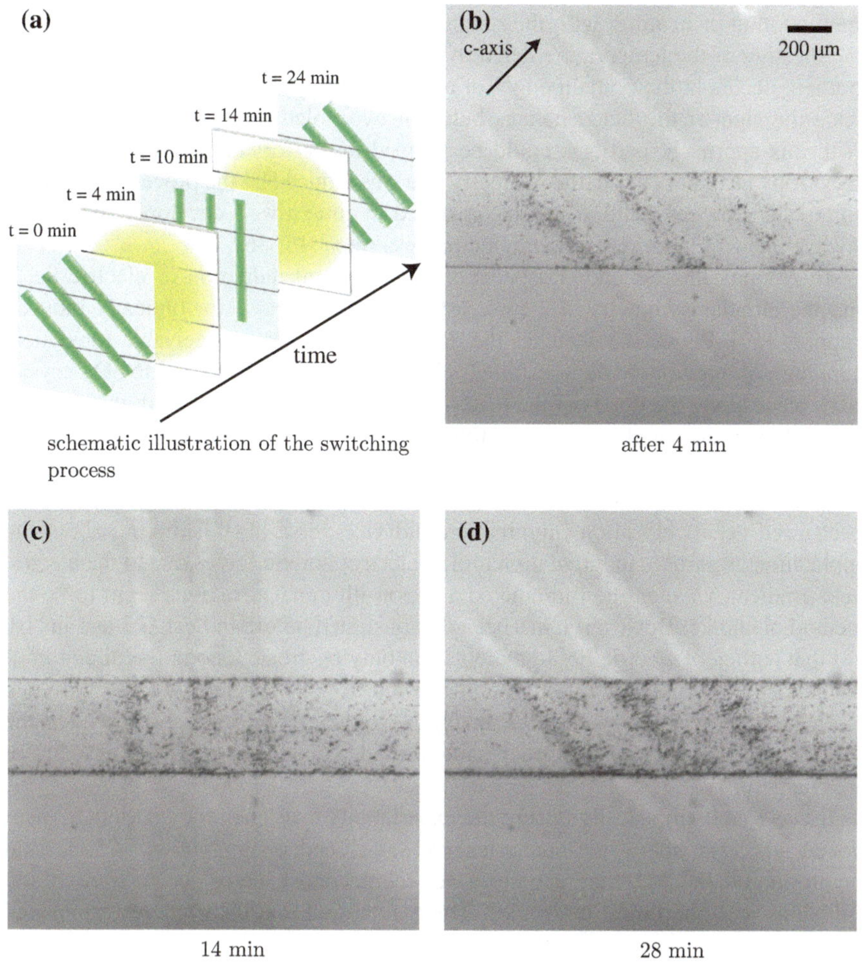

Fig. 6.2 Switching of trapping patterns in a microchannel: graphite particles can be trapped out of the flow by dielectrophoretic forces, which can be optically reconfigured in different configurations shown in (**b–d**); between each new induction steps, old gradients are erased by incandescent illumination as depicted in (**a**) (Images obtained by Stefan Gläsener, also published in [5])

axis [10–12]. By appropriately designing the trapping pattern or rotating the crystal, respectively, such problems may be reduced, but never be completely avoided. In order to demonstrate optically induced-trapping without anisotropy, one can employ a $LiNbO_3$ crystal in a z-cut geometry. Due to the fact that here the c-axis is normal to the crystal surface (cf. also Fig. 6.3), the geometry is inherently circular symmetric. Previous experiments have demonstrated dielectrophoretic trapping on such crystals by the exploitation of pyroelectric fields [13]. In a pyroelectric material, such as $LiNbO_3$, excess charges and resulting electric fields are generated on the end faces upon cooling or heating, with their sign determined by the direction of the c-axis and the sign of the temperature derivative [14]. Consequently, a sharp electric field gradient occurs at the domain edges of a periodically-poled lithium niobate crystal since the sign of the field changes between two poled regions of opposite sign. With this approach, particles could be trapped in any orientation without crystal anisotropy [13], but only in the geometry that was induced by the process of periodic poling. This process, although it has matured in quite a well-developed technique, involves several laborious lithographic process steps [15, 16].

To combine the beneficial aspects of an all-optical induction of electric fields and the circular symmetry, a z-cut crystal can still be used for optically-induced trapping [17]. Figure 6.3 sketches the different geometries conceivable for optical trapping. As previously demonstrated, charge separation inside a $LiNbO_3$ crystal takes place along the c-axis of the material. In the case of z-cut crystal, this leads to the agglomeration of charges on the top and bottom surfaces and the creation of an electric field in the illuminated regions only. Depending on the crystal orientation, these faces will be denoted as +c or −c face. By definition, the +c face of the crystal is charged negatively upon illumination and vice versa [14]. Thinking about the application as particle manipulation tool, dielectrophoretic forces due to the electric field gradient between the illuminated and non-illuminated regions seem to be the method of choice. However, it has been demonstrated before in Sect. 5.3 and in [10, 18] that sufficient electric field gradients can only be created along the direction of the c-axis of the material, but not perpendicular to it. This rules out the occurrence of dielectrophoresis in the z-cut geometry. Nevertheless, the charges present on the surface of the crystal can still attract matter provided that the particles in a suspension are charged.

It was found empirically during the experiments that when the previously mentioned glassy carbon microparticles were suspended in NOVEC 7500 fluid (hydrofluorether, 3M), they acquire a net charge, most likely due to friction and collisions with the liquid molecules. The assumption that the particles are indeed

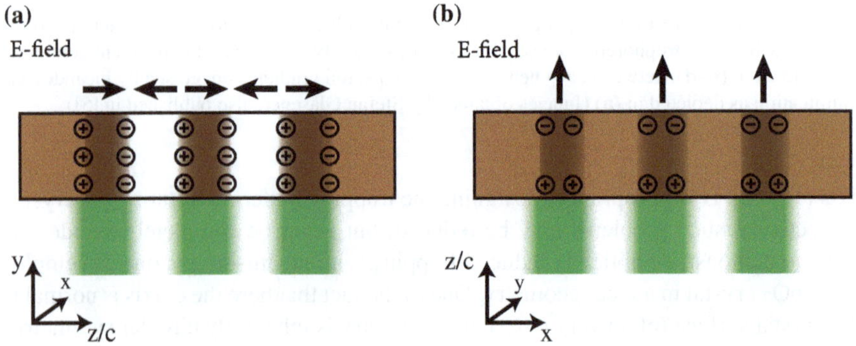

Fig. 6.3 Geometry and direction of electric fields for (**a**) y-cut (**b**) and z-cut optical trapping on a lithium niobate crystal; for z-cut crystals, field gradients are too low for dielectrophoresis, but accumulated charge carriers attract charged particles (also published in [17])

charged can be tested on a y-cut lithium niobate sample: if a sinusoidal electric field grating is present inside such a crystal, it has been shown that the field gradients are large enough for dielectrophoresis to occur. From the theoretical description of DEP, it is known that matter is attracted towards the regions of highest field intensity for the case of positive DEP. From Maxwell's equation $\nabla \varepsilon E = \varrho$, it is known that the charges responsible for the electric field possess a $\pi/2$ phase shift to the electric field. Thereby, it is possible to distinguish between dielectrophoretic and electrophoretic attraction in such a sinusoidal potential: Dielectrophoretic forces attract matter towards the regions of highest electric field intensity. On the other hand, electrophoresis guides particles into the regions where the charges responsible for the field are located.

Figure 6.4 illustrates the results of an experiment where a sinusoidal field (period $\Lambda = 420\,\mu\text{m}$) was inscribed into a y-cut lithium niobate crystal by an appropriate sinusoidal intensity modulation. The electric field was obtained by a DHM measurement and averaged along the constant direction as described in Sect. 3.6.3.2. After the electric field has been induced, the suspension of test particles is dripped onto the crystal surface. The same experiment is performed once with glassy carbon particles suspended in tetradecane, which is an apolar, non-dissociating oil, where dielectrophoretic forces are known to occur [10]. The second experiment follows the very same procedure but uses NOVEC 7500 as the suspension medium for the same particle type and same concentration of approximately $3\,\text{mg}\,\text{mL}^{-1}$. For both experiments, the particle concentration was derived by taking an inverted gray scale microscope image of the field-of-view. One might argue that there obviously is a cutoff in the particle density that can be detected, since particles lying on top of each other cannot be adequately resolved. However, the position of trapped particles with respect to the electric field pattern can still be accurately determined [17]. It is obvious that for the case of tetradecane as the suspension medium, the particles are attracted towards the regions of highest field intensity, just as it would be expected from the theoretical considerations in Chap. 2. For the case of NOVEC 7500 as the

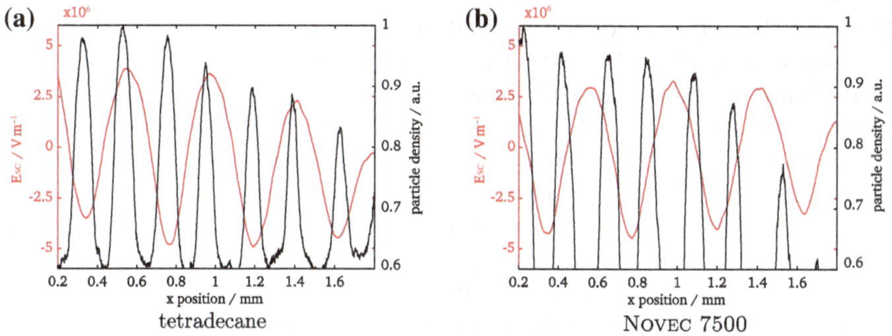

Fig. 6.4 Correlation of charge state and trapping position: **a** GC particles suspended in tetradecane are uncharged, hence attracted by DEP forces to regions of high field intensity; **b** GC particles in NOVEC 7500 are charged positively and negatively, hence being trapped by electrophoresis where the density of internal charges is highest (also published in [17])

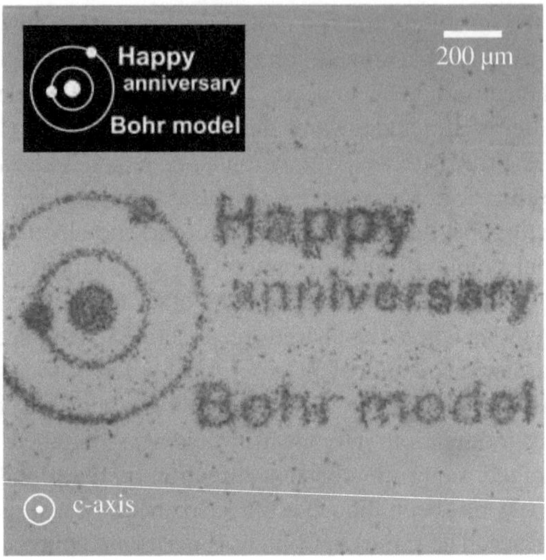

Fig. 6.5 Electrophoretic trapping of charged GC particles suspended in NOVEC 7500 in arbitrary patterns on a z-cut lithium niobate; inset shows the desired pattern, honoring the 100th anniversary of Bohr's semiclassical shell model (also published in [19])

suspension medium, there is a phase shift between electric field and trapped particles that is typical for electrophoretic attraction, as explained previously.

With these experiments, it has been proven that dielectrophoretic trapping dominates in a suspension of GC in tetradecane, whereas electrophoresis is the dominant mechanism if NOVEC 7500 is used as the suspension medium. This knowledge can be used to demonstrate optically-induced trapping in arbitrary pattern on a z-cut $LiNbO_3$ sample [17]. The z-cut crystal *142.3.2* for these experiments was again fabricated by the cooperation partners at the University of Padua and was iron-doped at concentration of $18.8 \times 10^{18}\mathrm{cm}^{-3}$ and a reduction degree of $c_{\mathrm{Fe}^{2+}}/c_{\mathrm{Fe}^{3+}} = 0.16 \pm 0.01$. Figure 6.5 shows the surface of the crystal after it has been irradiated with the amplitude pattern sketched in the inset for 10 s at a low mean intensity of $8.9\,\mathrm{m\,W\,cm}^{-2}$. As explained previously, in the illuminated regions charges accumulate that are able to attract charged matter of the opposite sign. The trapping pattern was chosen not only to honor the 100th anniversary of Niels Bohr's discovery of the semi-classical atomic shell model, but also because it contains all possible orientations of amplitude modulations. The results clearly shows that in the presented geometry, trapping of particles is possible in truly arbitrary patterns, circumventing previous restriction that originate from the crystal asymmetry.

Moreover, as stated before, the sign of the charges that accumulate on the illuminated surface depends on the orientation, with positive charges accumulating on the −c face of the crystal and vice versa. Especially, it can be deduced from the previous experiments shown in Fig. 6.4 that both negatively and positively charged particles exist in a suspension of GC in NOVEC 7500. This situation implies that

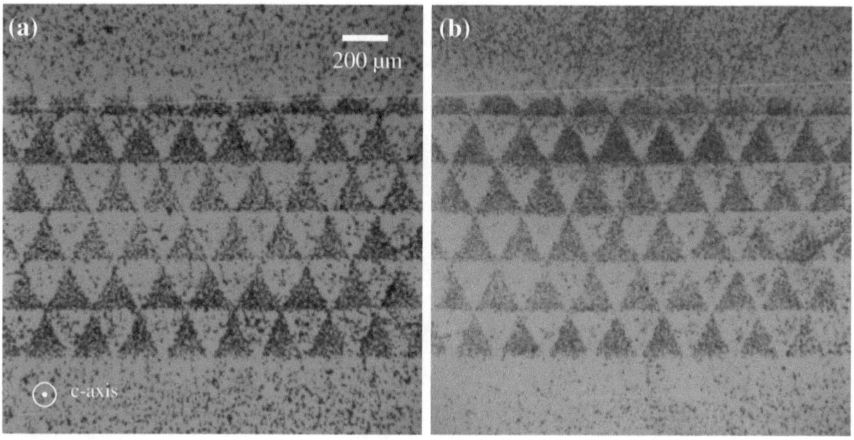

Fig. 6.6 Electrophoretic trapping on both faces of crystal 142.3.2; due to the fact that charged GC particles of both signs occur in a suspension with NOVEC 7500, on both sides of the crystal 50 % of the particles are attracted to the surface by electrophoresis. **a** +c face; **b** −c face

the electrophoretic attraction of particles is possible regardless of the orientation of the c-axis of the crystal. Figure 6.6 verifies that the same orientation of the trapped GC particles occurs, even when the crystal is turned by 180°. On the other hand, for the related substance NOVEC 7300 (hydrofluorether of different composition), it was found that GC beads suspended in this fluid acquired almost only positive charges, which means a different trapping behavior for both crystal orientations: since the +c face is charged negatively upon illumination, GC particles in NOVEC 7300 are strongly attracted towards illuminated regions on the +c face, whereas they are repelled from these region on the −c face due to the strong electrostatic repulsion between like charges. If at all within the field-of-view, they come to rest in the dark regions (see Fig. 6.7). A direct implication of these findings is that the knowledge about the crystal orientation allows the determination of the polarity of confined particles, which means that z-cut lithium niobate samples lend themselves as charge sensors to optofluidic applications.

6.3 Fabrication of Polymer Gratings on Photorefractive $LiNbO_3$

The micromanipulation capabilities of photorefractive crystals are not limited to the previously presented cases of a suspension of particles. The generality of derivation for dielectrophoretic forces has motivated unconventional applications like the structuring of polymer materials on the surface of a $LiNbO_3$ crystal [20]. In this case, a thin PDMS layer spin-coated onto the substrate represents the "particles" whereas air serves as the "suspension medium". Using this technique, it was possible to generate

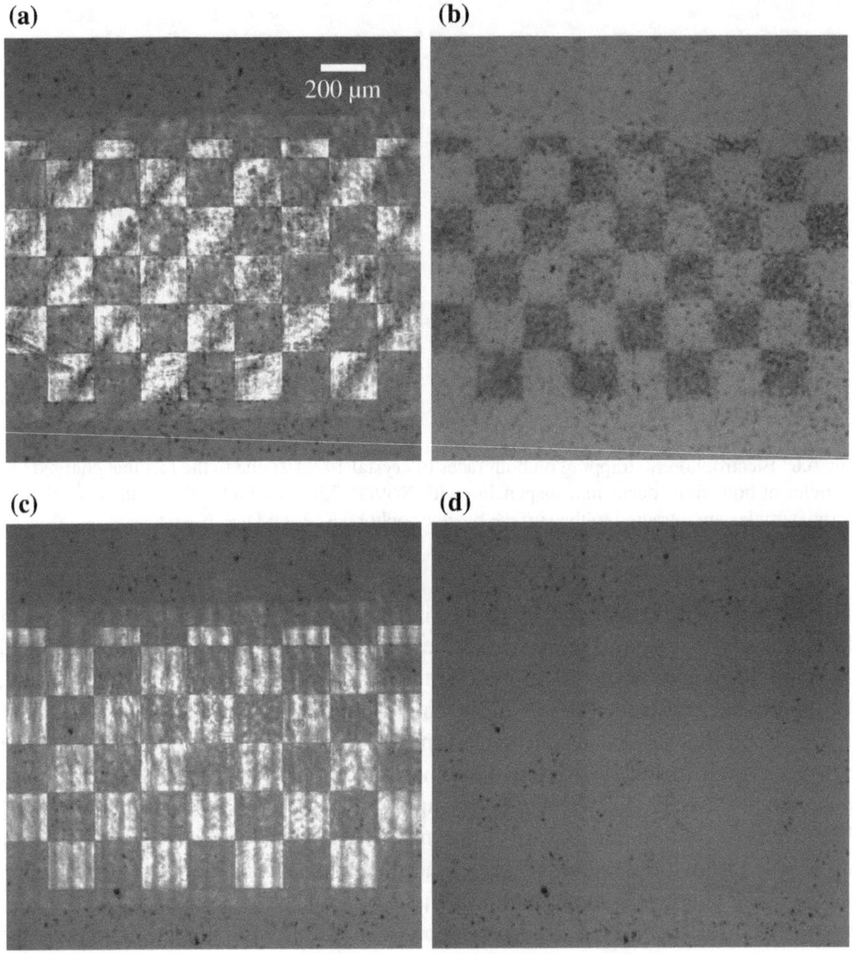

Fig. 6.7 Z-cut $LiNbO_3$ as a charge sensor: since the sign of accumulated charges depends on the orientation of the c-axis, positively charged GC beads are attracted to the negatively charged +c face, while they are repelled from the −c face, allowing to determine the charge sign from their trapping position (also published in [17]). +c face, incident pattern (**a**); +c face, trapped particles (**b**); −c face, incident pattern (**c**); −c face, repelled particles (**d**)

sinusoidal diffraction gratings as well two-dimensional patterns created by binary illumination [20, 21]. In this section, it will be demonstrated that the careful engineering of the illuminating light field enables the creation of an even larger variety of structures. The general approach is to spin-coat a thin PDMS layer on a photorefractive substrate and to illuminate this with a carefully engineered light field prior to curing the polymer. After solidification, the internal space-charge fields can be erased and the thickness of the modulated PDMS layer can be evaluated by digital holographic means.

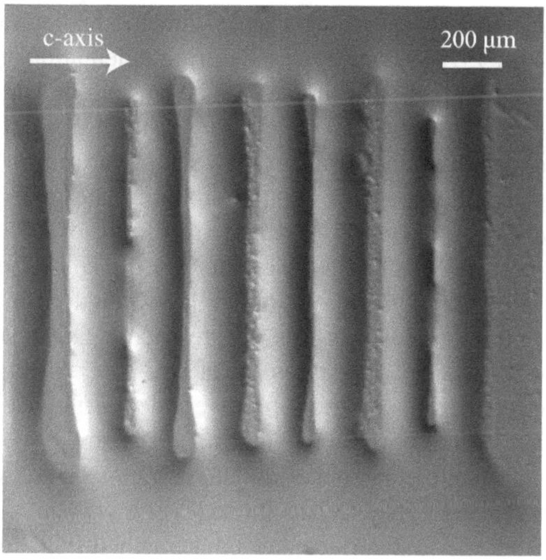

Fig. 6.8 PDMS grating on $LiNbO_3$; voids between the fringes are formed due to either too high forces or an insufficiently thick PDMS layer; due to many phase errors, the grating height in such a grating can only be mapped with DHM for single lines of PDMS. Image obtained by Daniel Chaparro González [29]

6.3.1 Thickness Measurement of Spin-Coated PDMS Layers

Before actually structuring a layer of PDMS on a $LiNbO_3$ substrate, it is indispensable to determine the thickness of the coated layers. Since the magnitude of DEP forces decays rapidly with increasing distance to the surface [22], the spin-coated layer must not be too thick. On the other hand, if a layer is too thin, a situation may arise where all the PDMS is removed from a particular part of the crystal (cf. Fig. 6.8), leaving behind the bare crystal with little dots of PDMS, a situation which is just as undesired since the aim is to produce a regular continuous diffraction grating. The thickness of a spin-coated layer depends on the rotation frequencies, the spin-coating time and the viscosity of the coating material [23]. For the determination of the thickness, a layer of PDMS is spin-coated onto a 500 μm thick $LiNbO_3$ sample and cross-linked for 45 min at 100 °C, as recommended by the polymer manufacturer's data sheet [24]. In general, there are a couple of measurement approaches that could be employed to determine the thickness of a such a layer, among them digital holography to determine the phase retardation by the PDMS layer. Knowing the material's refractive index, this allows the direct calculation of its height. In order for a DHM measurement to yield a reliable result, the phase retardation due to the layer should either be less than 2π or rise continuously from zero to its maximum value, so that phase ambiguities can be resolved by unwrapping the phase information. Due to the fact that ideally the whole surface of the substrate is covered with PDMS, such a reference point of $h_{PDMS} = 0$ cannot be found. As a different measurement technique, the use of profilometer has been proposed [25]. Due to the small Young's modulus of the material even in the solid state [26], these measurements have yielded smaller values than optical techniques because the PDMS is compressed during indentation. Therefore, it is the best way to obtain the thickness of the PDMS layer from an angular transmission spectrum.

Such a spectrum can be obtained in a very easy setup. The sample to be investigated is mounted on a high resolution rotation stage and illuminated with a collimated laser beam. In the present case, the rotation stage had a resolution of $\Delta\theta = 0.0006°$ and the laser was a helium-neon laser with $\lambda = 632.8$ nm. The transmitted power is measured by a powermeter while varying the rotation angle. The reason that the transmitted intensity varies with this angle is thin-film interference: when a coherent light wave is transmitted through a transparent slab of refractive index n_1, a part of the light is reflected inside the material due to the refractive index difference between the slab and the surrounding medium of refractive index n_0. The strength of this reflection in general depends on the angle of incidence and the refractive indices of both materials and can be calculated using Fresnel's Equations [27]. A portion of the reflected light is again reflected from the first surface and interferes with the directly transmitted light. The thickness of the slab determines whether the directly transmitted light and reflected light are in-phase, corresponding to a transmission maximum or π phase-shifted, corresponding to a relative transmission minimum. Taking into account all multiple reflections, one arrives at the well-known Airy formula [27]:

$$T(\delta) = \frac{1}{1 + F\sin^2(\delta)} \tag{6.1}$$

where T is the normalized transmission and $F = 4R/(1-R)^2$ is the finesse coefficient of a system with end face reflectivity R. Due to the fact that in this case, the reflection originates purely from Fresnel reflections on the surface, the reflectivity and hence the modulation of the transmission function is rather low. The parameter $\delta = 2\pi n_1 d \cos\theta/\lambda$ describes the optical path difference between the directly transmitted and reflected beam. From this parameter, it is evident that by scanning the angle of incidence for incoming light, relative transmission maxima and minima are obtained. Figure 6.9 shows the simulation of this angular spectrum.

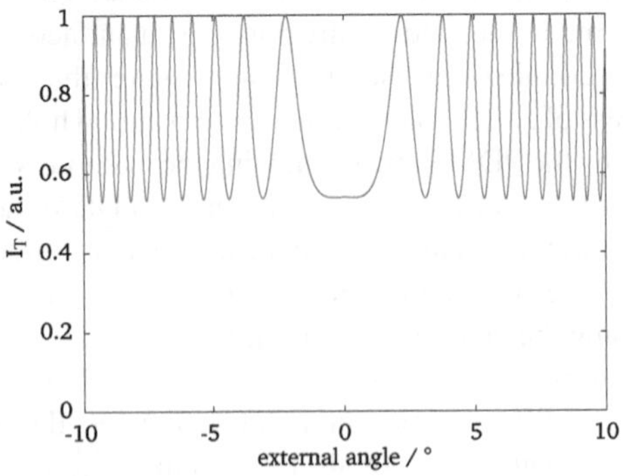

Fig. 6.9 Simulation of the angular transmission spectrum through a 500 μm thick $LiNbO_3$ slab

Having obtained such a spectrum in a real measurement, the thickness d of the object could be obtained by fitting the Airy function to the obtained data. However, the large number of free fitting parameters renders the fitting process error-prone. An easier and very robust way to obtain the thickness is to determine the angular position of all maxima or minima in the angular spectrum, respectively. Since these points correspond to constructive or destructive interference of both light beams, the respective Δ values are separated by an integer multiple of π, corresponding to the maxima or minima of the $\mathtt{sin}^2$ function and being equivalent to the condition

$$\Delta_k = \frac{2\pi}{\lambda} n_1 d \cos\theta_k = \pi k. \tag{6.2}$$

If the cosine values of the maxima/minima in the angular spectrum are plotted over their respective orders k, the thickness d can be obtained from the slope m of the linear fit to the data points:

$$\cos\theta_k = k \cdot m = k \cdot \frac{\lambda}{2n_1 d}$$
$$\rightarrow d = \frac{\lambda}{2n_1 m} \tag{6.3}$$

Note that for all of these calculations, the refractive index of the medium has to be known, since it is only possible to measure the optical thickness $n_1 d$ of the material with this technique. For all materials under investigation in this section, the refractive indices have been measured with sufficient precision. The method presented here was verified by measuring an undoped $LiNbO_3$ slab from a wafer, where it yielded within the measurement accuracy a thickness of 500 µm, the value specified by the vendor. So far, only a one-layer system surrounded by air has been considered. Since a thin layer of PDMS—which is the subject of interest in this section—alone lacks the mechanical stability to be measured, the angular transmission spectrum must always be measured in combination of PDMS/substrate two-layer system. To calculate the exact solution for the transmission function of such a system, it is necessary to take into account a coherent superposition of all internal reflections, which leads to the following reflection coefficient for the four-layer system of (air/PDMS/$LiNbO_3$/ air) [28]:

$$r_{\text{total}} = \frac{r_{12} + r_{234} e^{-2i\phi_2}}{1 + r_{12} r_{234} e^{-2i\phi_2}}, \tag{6.4}$$

with

$$r_{234} = \frac{r_{23} + r_{34} e^{-2i\phi_3}}{1 + r_{23} r_{34} e^{-2i\phi_3}} \tag{6.5}$$

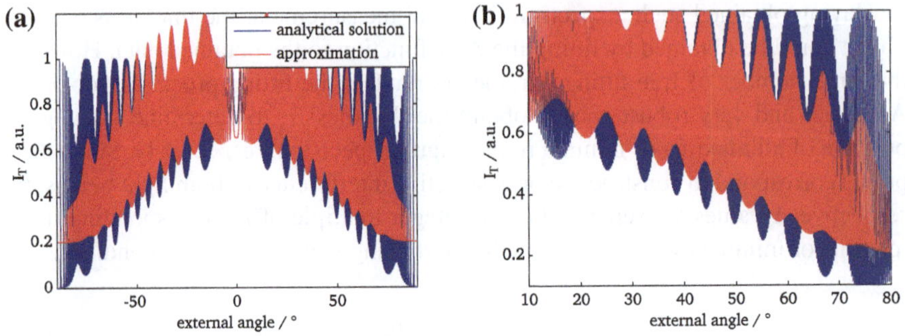

(a) numerical simulation and approximation

(b) close-up for direct comparison

Fig. 6.10 Simulation of the transmission function of **a** a two-layer system of 8 μ m PDMS on 500 μ m $LiNbO_3$; **b** close-up showing the coincidence of transmission maxima and minima for exact and approximate solution. Note that both curves have been offset with respect to each other for better visibility

as the reflection coefficient of the $LiNbO_3$ layer, sandwiched between air and PDMS. The r_{ik} in this case denote the reflection coefficients derived for each transition from the Fresnel equations. Although the analytical expression for the transmitted beam, which is derived from $T = 1 - r_{\text{total}}^2$, is quite complicated, an easy way to decouple the PDMS thickness from the rest of the sample can be deduced from the numerical simulation (cf. Fig. 6.10). Comparing the exact solution to an approximated semi-coherent solution, obtained by calculating the transmission functions for PDMS and $LiNbO_3$ independently and multiplying them, it becomes obvious that both functions contain regular maxima and minima that are only phase-shifted with respect to one another, but otherwise have the same periodicity. The high-frequency modulation in the angular spectrum originates primarily from the thick $LiNbO_3$ layer, whereas the low-frequency modulation is caused by the thinner PDMS layer, for which the transmission maxima have a higher angular separation. Since for the semi-coherent approach, it can be easily deduced that the low frequency minima in the semi-coherent transmission function occur at angular positions in the PDMS layer, for which the condition 6.2 is fulfilled, the same can be concluded for the minima in the exact transmission. The contributions from higher order reflections only slightly alter the form and absolute transmission of the transmission function, but obviously do not influence the separation of transmission maxima or minima. From the position of the secondary, low frequency maxima/minima, the thickness of the polymer layer can be obtained from a simple linear fit (cf. Fig. 6.11). Note that the results obtained for the thickness have been checked by cutting the PDMS layer in half and looking at it from the side. This procedure is less accurate than the optical transmission mode, which is the reason why it was not used in the first place, but serves as good double-check to verify the correctness of the thickness values.

Figure 6.12 shows the so-calculated thicknesses for a variety of rotation frequencies. For all rotation speeds, the total spin-coating time was 5 min since during the experiments longer spin-coating times were found to give more homogeneous layers.

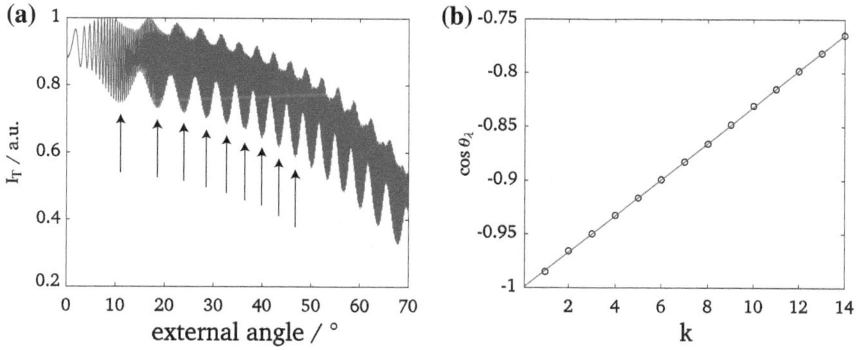

Fig. 6.11 Thickness measurement with the two-layer method: experimental transmission spectrum of PDMS spin-coated onto 500 μ m at 35 rps for 5 min; linear fit to the -cosine values of minima; from the linear slope, the thickness was determined to be 13.3 ± 0.1 μ m. **a** Angular transmission spectrum for two-layer system PDMS/$LiNbO_3$, **b** linear fit to minima in spectrum

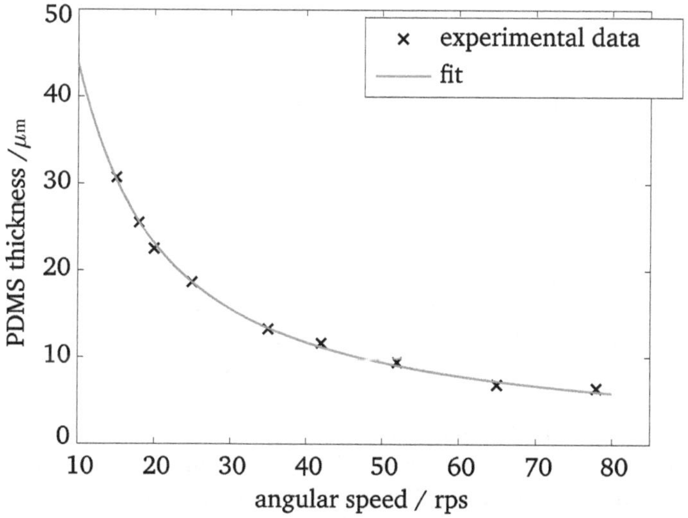

Fig. 6.12 Thickness of PDMS layer as derived by Eqs. 6.2–6.3 plotted over the spinning velocity; $t_{spin} = 300$ s for all samples; the solid curve is a fit to the model presented in Eq. 6.6

It was found for the set of parameters tested, the 5 min interval enables the creation from 31 down to 6.5 μ m thick layers, which was considered sufficient for the experiments in this section. The solid curve in Fig. 6.12 is a fit of the theoretical model about the spin-coating process of viscous liquids to the obtained data points [23]:

$$h(f,t) = \frac{h_0}{\sqrt{1 + c h_0^2 f^2 t}} \tag{6.6}$$

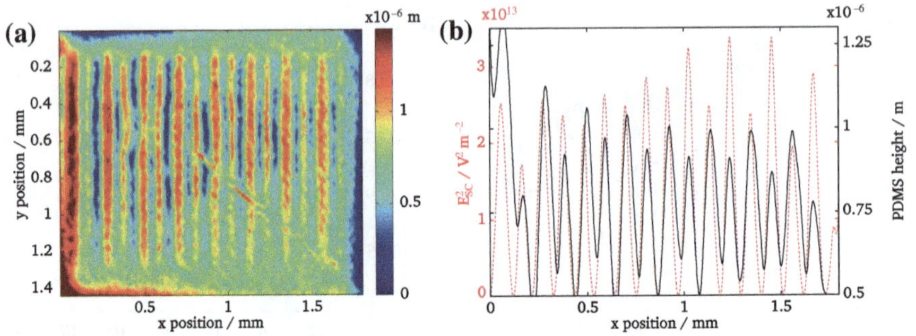

Fig. 6.13 Fabrication of PDMS gratings on photorefractive $LiNbO_3$; by inscribing a triangular internal field, the PDMS near the surface is moved by positive dielectrophoresis and forms a regular diffraction grating. Initial height of the PDMS layer was 7.7 μ m (Image obtained by Daniel Chaparro González [29]). **a** PDMS height map; **b** correlation of field intensity and PDMS height

where the spin-coating time $t = 300\,\mathrm{s}$ is fixed and the initial height $h_0 = 175\,\mu\,\mathrm{m}$ and material parameter $c = 1.538 \times 10^4\mathrm{s\,m^{-2}}$ are the fit parameters.

6.3.2 Optically-Induced Structuring of PDMS Layers

Based on the previous measurements, it is now possible to apply thin layers of PDMS of a desired thickness to the surface of a photorefractive $LiNbO_3$ crystal. Prior to curing the PDMS at elevated temperatures, the photorefractive surface is illuminated with an engineered light pattern. In order to be able to directly correlate the position of light pattern and the height of the PDMS layer after curing, a programmable heating plate is integrated into the combined optical structuring/phase measurement microscope setup that has been described and used before. With this setup, all process steps can be performed in a single position without having to move the crystal. Figure 6.13 shows the PDMS modulation resulting from illuminating the crystal with a triangular pattern with grating period of $\Lambda = 208\,\mu\,\mathrm{m}$. Contrary to the previous chapters, which have made use of binary patterns to create strong dielectrophoretic forces to trap very small particles, for the structuring of PDMS, binary patterns are not the pattern of choice. The field intensity gradients are very strong but only located to the regions of dark/bright interchanges in the binary illumination, so that PDMS is not manipulated everywhere. Due to this reason, illumination patterns were chosen that provide broader, but more homogeneous field intensity gradients inside the $LiNbO_3$ crystal, such as the triangular pattern used here.

Before the illumination of a 7.7 μ m PDMS film (spin-coated at 60 rps) with mean intensity of $1.72\,\mathrm{W\,cm^{-2}}$ for 2 min, the PDMS was pre-cured at 70 °C for 15 min. Pre-curing is done to increase the initial viscosity, which was found to be a good processing step to increase the smoothness of the resulting patterns. Without initial heating, the PDMS was often removed from certain areas of the crystal, similar to the situation depicted in Fig. 6.8. After structuring, the PDMS was cured for

1 h at 100 °C. The height of the PDMS can be retrieved from a DHM interferogram, provided that the space-charge field inside the $LiNbO_3$ samples is erased beforehand. Otherwise, a superposition of phase retardations from the refractive index modulation and the PDMS surface relief will lead to faulty results. In Fig. 6.13b, which shows the average of both electric field and PDMS modulation along the constant direction, it can be nicely seen that the previous assumptions about the dielectrophoretic force hold also true for the manipulation of continuous liquid polymer, namely that for liquids with a higher conductivity than that of the surrounding medium, air in this case, the PDMS is moved to regions of high field intensity. It has been claimed in previous publication about the reforming of PDMS on a $LiNbO_3$ substrate that PDMS "tends to concentrate across the illuminated regions, draining away from the dark regions" [21]. While this statement is correct for a single bright line inscribed into the crystal, for a regular periodic pattern, PDMS is attracted to dark and bright regions alike, since the field intensity is highest both in the dark and bright regions, hence leading to the characteristic frequency doubling observed in DEP manipulation [10, 18].

Of course, it is not sufficient to show that PDMS is modulated, but there should also be a functionality behind the structure. In Ref. [21], PDMS was cured to form the channel walls of a microchannel that was sealed on top by a glass slide. The structure from Fig. 6.13 due to its regular structure can be employed as a diffraction grating. From the separation of the PDMS maxima of $g = 106 \pm 6\,\mu$ m, the diffraction maxima of a regular grating are expected under an angle of $\theta = 0.341 \pm 0.02°$. Figure 6.14a shows the image of the diffraction pattern as seen on a CCD camera located $d = 21.5$ cm from the PDMS grating. The calculation of the angular diffraction spectrum from the pixel pitch and the separation d shows that as expected, the highest diffraction maximum occurs under an angle of $\theta_{exp} = 0.342 \pm 0.016°$. Additional maxima can be found under a smaller angle of $\pm0.185°$. These maxima

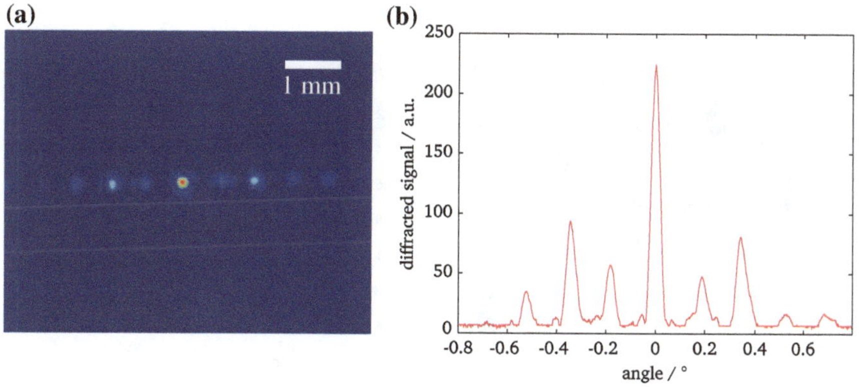

Fig. 6.14 Diffraction pattern from 1-dimensional PDMS grating; the measured diffraction angle of $0.342 \pm 0.016°$ matches nicely the expected angle of $0.341 \pm 0.02°$, calculated from the separation of the PDMS maxima (Image obtained by Daniel Chaparro González [29]). **a** diffraction pattern at d=21.5 cm; **b** angular diffraction spectrum

correspond to a PDMS grating of an approximate period of $g_2 = 196\,\mu$ m, and are due to the superstructure present in the PDMS pattern. This is roughly the period of the incident light pattern, and it is obvious, although the incident illumination was perfectly symmetric, the PDMS is attracted more towards regions, where the electric field is negative. The reasons for this are currently unknown and have to be investigated in further experiments, since from dielectrophoretic theory, such an asymmetry is not expected. Nevertheless, the good quality of the diffraction spots as observed in holds the promise that very flexible optical elements can be fabricated out of PDMS pre-structured and cured on $LiNbO_3$ substrates.

Apart from the fabrication of sinusoidal PDMS fringes, the optical flexibility enables further interesting shapes, two of which are illustrated in Fig. 6.15. The left image shows the close-up of PDMS fringes that were formed by illuminating a $LiNbO_3$ crystal with a one-dimensional stepwise exponential intensity pattern, which can be described by

$$I(x) = I_0 e^{\frac{mod(x,\Lambda_x)}{\sigma x}} ,$$

where I_0 is the maximum intensity, x the pixel index, Λ_x the periodicity, and σ_x a scaling factors that varies the steepness of the exponential rise. Such a pattern is inherently asymmetric and different gradients on both sides lead to the formation of almost triangular PDMS fringes. A grating of such form is typically denoted as a blazed grating and has the ability to diffract incoming light primarily into a single angular direction. Figure 6.15b shows a two-dimensional sinusoidal pattern, obtained by structuring the underlying $LiNbO_3$ crystal with a two-dimensional triangular pattern, analogous to that used in Fig. 6.14. The use of such a grating is obviously to diffract light in a discrete regular pattern, the angular separation of the diffraction spots determined by the lattice constant. Although the image contains some unwrapping errors, the general structure of the pattern can clearly seen. These errors result from points in the pattern, where the PDMS was completely removed from the crys-

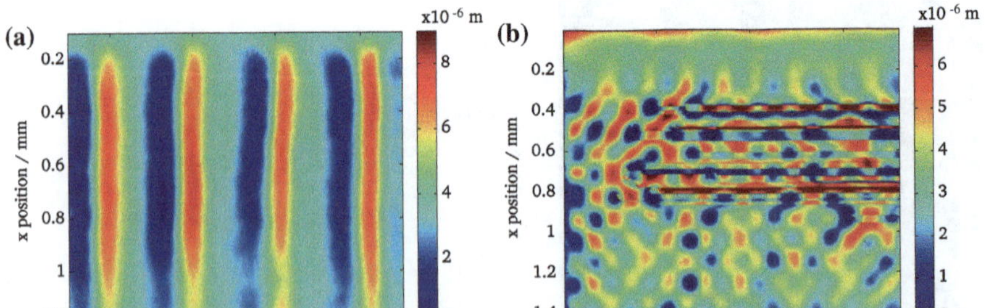

Fig. 6.15 Advanced structures fabricated by PDMS curing on pre-structured $LiNbO_3$: **a** asymmetric ratchet pattern, which forms on crystals illuminated with a step-wise exponential light pattern; **b** two-dimensional diffraction grating formed on a crystal illuminated by a 2D triangular pattern, similar to that of Fig. 6.13; note that (**b**) contains some unwrapping errors, for details refer to text (Images obtained by Daniel Chaparro González [29])

tal, and the phase could not be correctly calculated because the interference pattern is not continuous. Even the most robust phase unwrapping algorithms that were used in this thesis (see also Appendix A) failed in unwrapping such a pattern.

6.4 Optofluidic Router

The last demonstrator concept is concerned with optofluidic applications of photorefractive substrates. Due to their crystalline nature, the photorefractive substrates used in this thesis are inherently chemically resistant and therefore can be exposed to a wide variety of substances. Since they were originally made for optical applications, their optical quality is comparable to or even better than most microfluidic parts and it has been shown in Sect. 6.1 that $LiNbO_3$ crystals can be used as bottom of a PDMS microchannel. While in the previous application, the particle trapping out of microfluidic stream was the subject of interest, this section will demonstrate that electric field gradients can also manipulate fluids in fluids. In particular, this refers to microfluidic droplets that become more and more important, since they enable a continuous and controlled synthesis of CdS nanoparticles [30], polymer microparticles of different shapes [31] or Janus particles with two separate sides that feature different optical and electrical properties [32, 33].

6.4.1 Droplet Generator Design

Two different droplet generators were fabricated by PDMS soft lithography, which is explained in more detail in Appendix B. Their geometries are presented in Fig. 6.16. In general, the concept of droplet generation requires two immiscible fluids, one of which is supplied in a high volume base stream into which droplets are injected. The

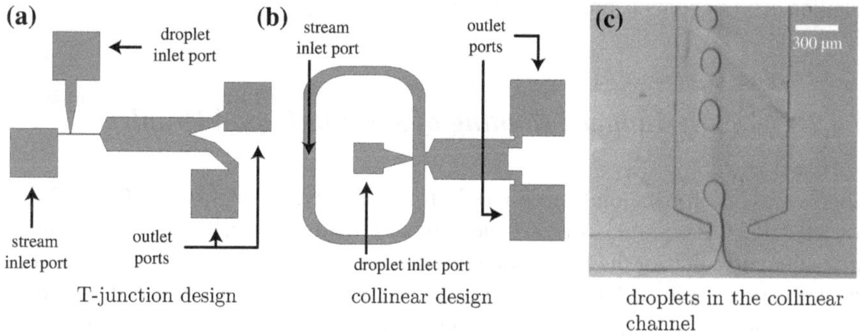

Fig. 6.16 Different designs for droplet generators; T-junction shaped geometries (**a**) possess the advantage of being easy to connect and control, but droplets strongly adhere to the channel walls; in the collinear design (**b**), droplets initially have no contact to the side walls, which facilitates the fabrication of free-flowing droplets (**c**)[submitted to Laser and Photonics Reviews]

first geometry that was fabricated is the T-junction geometry, where the base stream is intersected by a droplet nozzle, so that droplets are cut off from the nozzle [34]. This geometry has the advantage of being fairly easy to implement with only two necessary connections for the base liquid and the droplet liquid. Additionally, the fraction between droplets and base liquids can be easily varied by changing the flow speeds. However, it was found empirically that due to the production of the droplets in immediate contact to the channel walls, they have a tendency to stick to these walls, even if the channel gets broader (cf. Fig. 6.16a). While this does not pose a problem in narrow channel networks [35], the adhesion forces can be comparable to the strength of the dielectrophoretic forces, which impedes the droplet manipulation. The second design that can circumvent the problem of sticking droplets is the so-called *flow-focusing* geometry. In this case, three streams flow collinearly, with the base streams flowing at a higher flow rate so that the middle stream is efficiently segmented into discrete droplets [36]. It provides the advantage that the droplets are free-flowing and ideally do not have contact to the channel walls, but requires three independent streams of liquid, two for the base stream from each side and a central one for the the injection of droplets, which is harder to realize in limited space and requires one additional connection and pumping device. Note that depending on the size of the droplets, they may come into contact with the bottom and top sealing of the channels.

As a trade-off between both geometries, the collinear geometry of Fig. 6.16b was conceived. The base stream is connected on the left side of the device, with special care to realize a central connection so that there are inherently two equal flows to both sides. The droplet injection nozzle is connection via the square reservoir in the middle of the structure. Although this geometry does not provide the possibility to adjust the flow rates for the two sides of the focusing flow independently, it was found in the experiments that single droplets were generated sufficiently well. The main advantage of this design is that one of the additional pumping devices is economized, which could also have potential applications for Point-of-Care (PoC) microfluidic devices in developing world countries, for which the high cost of the ancillary components is one the main restrictions [37].

6.4.2 Optically-Induced Routing of Air and Liquid Droplets

The expression for the dielectrophoretic force, that was calculated in Chap. 2, has revealed that it is possible to attract and to repulse matter, based on their electric properties, in particular for the low-frequency case, their electrical conductivities. While the derivation was based on the assumption that only dipolar terms are induced in the sphere under investigation, the size of typical droplets in the current experiments is of the same size as the virtual electrode (cf. Fig. 6.16c). However, it has been shown that even in the presence of multipolar moments, the Clausius-Mossotti factor for higher order moments does not change its sign [38]. Also, if the Kelvin polarization force density is taken into account for continuous liquids, the NOVEC fluids are more

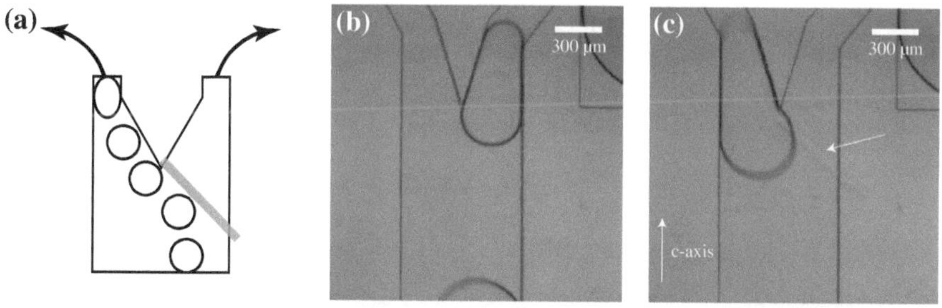

Fig. 6.17 Bubble sorting by virtual electrodes; without virtual electrodes in the substrate material, air bubbles exit the channel through the right outlet; after the induction of a binary stripe, the bubbles are reliably guided to the left port by negative dielectrophoretic forces; virtual electrode position is indicated by the white arrow [submitted to Laser and Photonics Reviews]. **a** schematic illustration; **b** before blocking outlet; **c** after writing virtual blocking electrode

polarizable than the tetradecane, with $\varepsilon_{TD} \approx 2$ and $\varepsilon_{7300/7500} = 6.14/5.83$ [39, 40]. It is therefore a safe assumption that even in the case of comparably large droplets, the forces acting on the droplet suspended in a base medium will be attractive for NOVEC droplets in tetradecane and repulsive if air or another less polarizable bubble is immersed by NOVEC fluid. Therefore, by constructing a droplet generating microchannel with a $LiNbO_3$ bottom layer and by inducing virtual electrodes, it is possible to steer and immobilize droplets and gas bubbles in a microchannel.

Figure 6.17 shows the results of a typical routing experiment. Air bubbles are generated in a NOVEC 7500 flow (hydrofluorether, 3M, $\sigma \approx 5.6 \times 10^{-12}\,\mathrm{S\,cm^{-1}}$) in the T-junction shaped geometry. The flow rate ratio between base and droplet flow in this experiment was adjusted to be 4:1. As explained previously, droplets or bubbles in this geometry have a tendency to stick to the side walls of the channel, which results in the bubbles exiting the device to the right outlet. Since air inherently has both a lower conductivity and dielectric permittivity, the forces generated by an electric field gradient will definitely be repulsive. This result can be used to route and steer the gas bubbles. Figure 6.17 shows that the right exit port can be closed by inscribing a virtual electrode that repels the air bubbles. After writing with a binary stripe pattern at $I = 0.56\,\mathrm{W\,cm^{-2}}$ for 12 s , all bubbles are reliably guided towards the left outlet port. This intensity may seem quite large at first, but considering that the total size of the barrier is small, the total power necessary for the inscription is only 0.14 m W, so by using a more efficient light modulating device than an ASLM, the writing time can be further reduced. Of course, the repulsive force acts not only to the side but also upwards. However, the air bubbles cannot travel *over* the *virtual barrier* due to the small height of the micro channel of 65 μ m.

By an appropriate selection of materials, it is also possible to generate the opposite force direction. Positive dielectrophoretic forces have been employed in Fig. 6.18, where droplets of NOVEC 7300 have been generated in the collinear flow droplet device in a base stream of tetradecane. Since both the conductivity as well as the dielectric permittivity of NOVEC 7300 is larger than that of tetradecane ($\sigma_{TD} \leq$

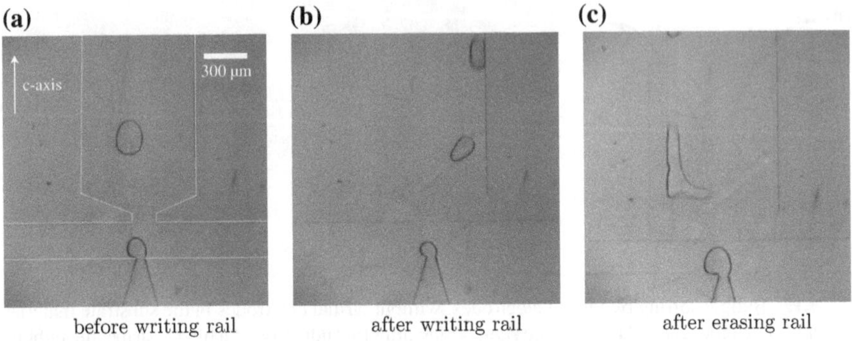

Fig. 6.18 *Virtual rails* for liquid routing in the collinear channel (highlighted by white line); on unmodified $LiNbO_3$ substrate *151.7*, droplets of NOVEC 7300 in a stream of TD **a** flow freely towards the exit; **b** after inducing *virtual rails* on the crystal surface, the attractive dielectrophoretic forces guide the NOVEC droplets towards the right exit port; **c** using the incandescent microscope illumination, the virtual rails can be erased, resulting in droplet flow to the left [submitted to Laser and Photonics Reviews]

$1 \times 10^{-15}\,\mathrm{Scm}^{-1}$), the Clausius-Mossotti factor will be larger than zero, which means that the NOVEC droplets are attracted by field gradients into regions of high field intensity. Qualitatively, the same results would apply, if the Kelvin polarization force density would be considered. The flow rates for a good generation of droplets were $5\,\mu\mathrm{Lmin}^{-1}$ for the tetradecane base flow and $0.4\,\mu\mathrm{Lmin}^{-1}$ for the NOVEC 7300 droplet infusion. Since the forces are now attractive, the previous virtual barrier becomes a *virtual rail* now, on which droplets are steered. The virtual rail was inscribed into $LiNbO_3$ crystal *151.7* at an intensity of $0.4\,\mathrm{W\,cm}^{-2}$ for approximately 15 s. Before the modification of the $LiNbO_3$ substrate, the droplets were flowing freely towards the exit ports, remaining more or less in the middle of the channel. After the inscription, droplets at the flow focusing nozzle follow the induced field gradient and are directed to the right exit. When a droplet reaches the end of the virtual rail, it may even be immobilized by the strong attractive forces, if it is not too large. For larger droplets, the Stokes friction that pushes the droplets forward increases linearly with the radius. Accordingly, it could be seen in the experiments that even if a droplet was immobilized at the end, it was removed by the next arriving droplet and either the whole droplet or most of its volume was released towards the exit port on the right.

As in the case of dielectrophoretically trapped particles in a microchannel, the induced electric field modulation can be erased using homogeneous illumination. Figure 6.18c shows the droplet movement after the virtual rail has been erased with the incandescent microscope illumination, which has the advantage of a broadband spectrum, hence no speckle noise. In principal, the same effect could be achieved with homogeneous laser illumination. The reason this light source is not used is only due to aesthetic reasons, since speckles always introduce a little modulation which is directly visible in the refractive index pattern. For the experiment in Fig. 6.18c, it

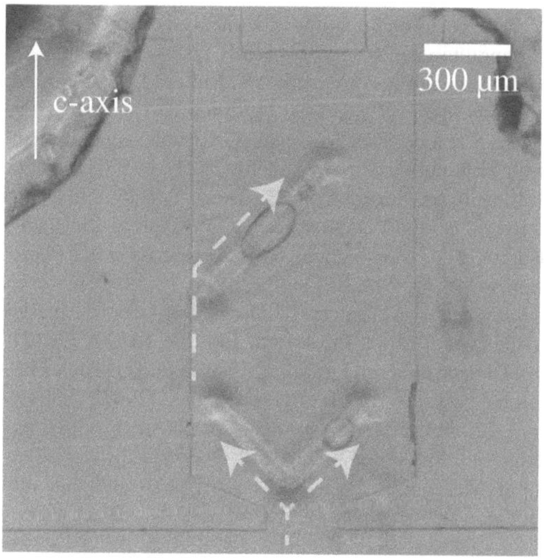

Fig. 6.19 Droplet splitting and complex routing on crystal *AT2*; NOVEC 7500 droplets in TD are split up at the "v" near the droplet nozzle, leaving them two possible ways indicated by the arrows; after the split-up, the droplet can be re-routed to the right side by the second virtual rail; note that this image has been acquired in Zernike phase-contrast mode to visualize the electrode structures [submitted to Laser and Photonics Reviews]

took 45 s for the erasure of the modulation, after which the droplets are not guided to the right exit port any more, but move towards the left exit. Note that this movement is not caused by any internal electric fields any more, but is most likely the result of an obstruction in the right channel or a biased flow. Even before the modification of the $LiNbO_3$ substrate, the droplet flow was slightly biased to the left. Of course, more complicated structures, such as a droplet splitter or a droplet maze can easily be induced thanks to the ASLM and its ability create arbitrary patterns (cf. Fig. 6.19).

References

1. G. Whitesides, The origins and the future of microfluidics. Nature **442**, 368–373 (2006)
2. D. Psaltis, S. Quake, C. Yang, Developing optofluidic technology through the fusion of microfluidics and optics. Nature **442**, 381–386 (2006)
3. C. Monat, P. Domachuk, B.J. Eggleton, Integrated optofluidics: a new river of light. Nat. Photonics **1**, 106–114 (2007)
4. J. Imbrock, S. Wevering, K. Buse, E. Kratzig, Nonvolatile holographic storage in photorefractive lithium tantalate crystals with laser pulses. J. Opt. Soc. Am. B: Opt. Phys. **16**(9), 1392–1397 (1999)
5. S. Glaesener, M. Esseling, C. Denz, Multiplexing and switching of virtual electrodes in optoelectronic tweezers based on lithium niobate. Opt. Lett. **37**(18), 3744–3746 (2012)
6. Y. Berdichevsky, J. Khandurina, A. Guttman, Y. Lo, UV/ozone modification of poly (dimethylsiloxane) microfluidic channels. Sens. Actuators, B: Chem. **97**(2–3), 402–408 (2004)

7. R. Kretschmer, W. Fritzsche, Pearl chain formation of nanoparticles in microelectrode gaps by dielectrophoresis. Langmuir **20**(26), 11797–11801 (2004)
8. A.L. Briseno, S.C.B. Mannsfeld, M.M. Ling, S. Liu et al., Patterning organic single-crystal transistor arrays. Nature **444**(7121), 913–917 (2006)
9. S.-K. Chae, C.-H. Lee, S.H. Lee, T.-S. Kim et al., Oil droplet generation in PDMS microchannel using an amphiphilic continuous phase. Lab Chip **9**(13), 1957–1961 (2009)
10. M. Esseling, F. Holtmann, M. Woerdemann, C. Denz, Two-dimensional dielectrophoretic particle trapping in a hybrid crystal/PDMS-system. Opt. Express **18**(16), 17404–17411 (2010)
11. J. Villarroel, H. Burgos, A. Garcia-Cabanes, M. Carrascosa et al., Photovoltaic versus optical tweezers. Opt. Express **19**(24), 24320–24330 (2011)
12. M. Esseling, S. Glaesener, F. Volonteri, C. Denz, Opto-electric particle manipulation on a bismuth silicon oxide crystal. Appl. Phys. Lett. **100**(16), 161903 (2012)
13. S. Grilli, P. Ferraro, Dielectrophoretic trapping of suspended particles by selective pyroelectric effect in lithium niobate crystals. Appl. Phys. Lett. **92**, 232902 (2008)
14. R. Weis, T. Gaylord, Lithium-niobate - summary of physical properties and crystal-structure. Appl. Phys. A Mater. Sci. Process. **37**(4), 191–203 (1985)
15. L. Myers, R. Eckardt, M. Fejer, R. Byer et al., Quasi-phase-matched optical parametric oscillators in bulk periodically poled $LiNbO_3$. J. Opt. Soc. Am. B: Opt. Phys. **12**(11), 2102–2116 (1995)
16. R. Batchko, V. Shur, M. Fejer, R. Byer, Backswitch poling in lithium niobate for high-fidelity domain patterning and efficient blue light generation. Appl. Phys. Lett. **75**(12), 1673–1675 (1999)
17. M. Esseling, A. Zaltron, C. Sada, C. Denz, Charge sensor and particle trap based on z-cut lithium niobate. Appl. Phys. Lett. **103**, 061115 (2013)
18. H.A. Eggert, F.Y. Kuhnert, K. Buse, J.R. Adleman et al., Trapping of dielectric particles with light-induced space-charge fields. Appl. Phys. Lett. **90**, 241909 (2007)
19. M. Esseling, A. Zaltron, N. Argiolas, G. Nava et al., Highly reduced iron-doped lithium niobate for opto-electronic tweezers. Appl. Phys. B Lasers Opt. **113**(2), 191–197 (2013)
20. L. Miccio, M. Paturzo, A. Finizio, P. Ferraro, Light induced patterning of poly (dimethylsiloxane) microstructures. Opt. Express **18**(11), 10947–10955 (2010)
21. L. Miccio, P. Memmolo, S. Grilli, P. Ferraro, All-optical microfluidic chips for reconfigurable dielectrophoretic trapping through SLM light induced patterning. Lab Chip **12**(21), 4449–4454 (2012)
22. J.K. Valley, A. Jamshidi, A.T. Ohta, H.-Y. Hsu et al., Operational regimes and physics present in optoelectronic tweezers. J. Microelectromech. Syst. **17**(2), 342–350 (2008)
23. A. Emslie, F. Bonner, L. Peck, Flow of a viscous liquid on a rotating disk. J. Appl. Phys. **29**(5), 858–862 (1958)
24. Dow Corning, Product information: Sylgard 184 silicone elastomer (2013)
25. W. Zhang, G. Ferguson, S. Tatic-Lucic, Elastomer-supported cold welding for room temperature wafer-level bonding. In *17th IEEE International Conference on Micro Electro Mechanical Systems Technical Digest*, (2004), pp. 741–744
26. W. Zhang, J. Labukas, S. Tatic-Lucic, L. Larson et al., Novel room-temperature first-level packaging process for microscale devices. Sens. Actuators, A **123–124**, 646–654 (2005)
27. E. Hecht, *Optik* (Oldenbourg, München, 2002)
28. P. Yeh, *Optical Waves in Layered Media* (Wiley, New York, 2005)
29. D.C. González, Master's thesis, Fabrication of polymeric diffraction gratings by means of optically induced dielectrophoresis, University of Münster, 2014
30. I. Shestopalov, J. Tice, R. Ismagilov, Multi-step synthesis of nanoparticles performed on millisecond time scale in a microfluidic droplet-based system. Lab Chip **4**(4), 316–321 (2004)
31. Z. Nie, S. Xu, M. Seo, P. Lewis et al., Polymer particles with various shapes and morphologies produced in continuous microfluidic reactors. J. Am. Chem. Soc. **127**(22), 8058–8063 (2005)
32. Z. Nie, W. Li, M. Seo, S. Xu et al., Janus and ternary particles generated by microfluidic synthesis: design, synthesis, and self-assembly. J. Am. Chem. Soc. **128**(29), 9408–9412 (2006)

33. T. Nisisako, T. Torii, T. Takahashi, Y. Takizawa, Synthesis of monodisperse bicolored janus particles with electrical anisotropy using a microfluidic co-flow system. Adv. Mater. **18**(9), 1152+ (2006)
34. T. Nisisako, T. Torii, T. Higuchi, Droplet formation in a microchannel network. Lab Chip **2**(1), 24–26 (2002)
35. K.R. Strehle, D. Cialla, P. Roesch, T. Henkel et al., A reproducible surface-enhanced Raman spectroscopy approach. Online SERS measurements in a segmented microfluidic system. Anal. Chem. **79**(4), 1542–1547 (2007)
36. S. Anna, N. Bontoux, H. Stone, Formation of dispersions using "flow focusing" in microchannels. Appl. Phys. Lett. **82**(3), 364–366 (2003)
37. B. Weigl, G. Domingo, P. LaBarre, J. Gerlach, Towards non- and minimally instrumented, microfluidics-based diagnostic devices. Lab Chip **8**(12), 1999–2014 (2008)
38. T. Jones, M. Washizu, Multipolar dielectrophoretic and electrorotation theory. J. Electrostat. **37**(1–2), 121–134 (1996)
39. 3M Deutschland GmbH. "Novec 7300 datasheet" (2013)
40. 3M Deutschland GmbH. "Novec 7500 datasheet" (2013)

Chapter 7
Summary

7.1 Conclusion

In this thesis, the general properties of photorefractive optoelectronic tweezers as substrate materials for the manipulation of micro particles and fluids were described. The main mechanism of particle alignment in these devices is based on dielectrophoretic forces, which only occur in the presence of high electric field gradients. Apart from static electrode arrays, such field gradients can also be created all-optically by photorefractive crystals. From the theoretical description of the charge transport processes, it was derived that only drift in an external field and the bulk photovoltaic effect (PVE) provide sufficiently high dielectrophoretic forces. The effect of diffusion could be ruled out due to the fact its magnitude in general is at least one order lower that for typical drift scenarios. One representative of each material class has been presented and thoroughly investigated, namely bismuth silicon oxide (BSO) for drift- and lithium niobate ($LiNbO_3$) for PVE-dominated materials.

The mere creation of electric fields, which could in principle be achieved in every photoconductor, is already sufficient for dielectrophoretic forces to occur. However, a very useful property that is only found in photorefractive materials is their inherent change of the refractive index with the electric field strength, the so-called *linear electro-optic* or *Pockels effect*. This effect enabled direct optical access to the electric field distribution by measuring the phase shift and hence the internal electric field using three different phase sensitive techniques. It was shown that the first method of measuring the diffraction efficiency and the inversion of the Kogelnik formula enables a fast calculation of the internal space-charge fields without any post processing apart from very simple arithmetic calculations. On the other hand, spatial information can only be gained by two-dimensional image acquisition techniques, of which Zernike phase-contrast is a standard and robust technique available in any scientific microscope. However, it does only yield limited quantitative information. As a combination between the beneficial aspects of both previous techniques, digital holographic microscopy has been introduced as the technique of choice due to its ability to generate quantitative electric field data in two-dimensions. It could be

M. Esseling, *Photorefractive Optoelectronic Tweezers and Their Applications*,
Springer Theses, DOI 10.1007/978-3-319-09318-5_7

shown that the imaginary part of the signal, which is necessary to calculate the phase retardation of the light wave, is a natural byproduct of the Fourier filtering process that is performed to suppress the carrier frequency of the interference pattern. In this way, even today a typical DHM image can be calculated in approximately 500 ms, for non-specialized processors and with non-optimized scripting languages such as MATLAB.

The technique of DHM was employed as versatile tool for the investigation of the relevant material properties. For $LiNbO_3$, it is well-known from previous publications that the internal fields in this material are exceptionally high. The main backdraft, however, is the slow speed at which these fields are formed. In order to speed up this process, a setup was realized in which the refractive index could be measured directly while at the same time the available light for writing the process was concentrated only on the desired active area of the crystal. This higher intensity directly increases the density of free charge carriers and hence the speed at which field are generated. In addition, in a cooperative project with the University of Padua, the material's photoconductivity was increased by a thermal reduction treatment. This treatment alters the ratio between the two valence states Fe^{2+} and Fe^{3+} of the iron dopant, which was introduced to enhance the photosensitivity. It was found experimentally that for reduction degrees higher than 1, for which a long thermal treatment is necessary, a disproportionate decrease of the saturation field sets it. The origin of this previously unknown effect can be attributed to the occurrence of intrinsic polaron or bipolaron defects, which have a strong influence on the sample conductivity, or the incorporation of hydrogen during the reduction treatment. The results of the respective section allows the selection of an intermediate reduction degree of less than unity, which already provides a more than tenfold increase in the writing speed while the saturation field is still sufficiently high for dielectrophoretic trapping.

In contrast to $LiNbO_3$, bismuth silicon oxide samples benefit from their inherently high photoconductivity which enables near real-time induction of internal fields at very low light powers in the range of $100\,\mu W\,cm^{-2}$. The challenging aspect of using BSO as substrate material for photorefractive optoelectronic tweezers is the fact that—as for all drift-dominated materials—the field strength is limited to the magnitude of the externally applied field E_0. Over and above, the internal field is further screened when the BSO sample is not fully illuminated, which, as a matter of fact, is a typical scenario in optical trapping experiments. By virtue of an alternating external voltage, the occurrence of screening effects could be suppressed and the internal field could even be enhanced above its external value. The theoretically predicted phase-shift between incident intensity pattern and resulting space-charge field in AC-biased BSO could be directly shown by DHM for the first time.

The results of these fundamental investigations were used to demonstrate dielectrophoretic as well as electrophoretic trapping on the surface of photorefractive crystal substrates. In accordance with the previous findings, no dielectrophoresis was observed on DC-biased BSO crystal due to the screening of the external field from the illuminated region, whereas in the AC-biased case, a modulation of the particle density was clearly visible. For highly reduced lithium niobate, a significant decrease in the trapping efficiency and contrast could be observed, likewise caused by the decrease in the field modulation for higher reduction degrees. Since in both material classes, the charge transport is inherently asymmetric, either caused by the direction of the applied field or photovoltaic axis, the selection of trapping patterns is reduced to a maximum angle of 60° and between the k-vector of the modulation and the c-axis of the crystal (for $LiNbO_3$, 45° for BSO). It was further investigated, how the suspension medium affects the strength of dielectrophoretic forces. In this thesis and in all previous publications, dielectrophoresis on photorefractive samples has only been demonstrated with particles suspended in highly insulating liquids. To describe the charge transport processes near the surface of the crystal, an equivalent circuit model was developed. Using this model, it could be predicted that sufficiently high fields for DEP can only be expected, if the suspension medium has a lower or comparable conductivity than the photorefractive substrate. Otherwise, the electric field near the surface will be discharged before any interaction with matter can take place.

The last and largest part of this thesis describes possible optofluidic applications for photorefractive tweezers. Optofluidics describes the merging and mutual influence of optics and microfluidics. In general, photorefractive crystals are well-suited as active substrate materials for microfluidic applications due to their chemical inertness and excellent optical quality. On top of this, $LiNbO_3$ benefits from the fact that is does not need an external voltage supply and hence can be easily integrated. Thereby, micro particles were trapped out of a PDMS micro channel, which was applied to photorefractive $LiNbO_3$ serving as the active bottom layer to trap matter and to seal the channel. It was further shown how the correlation of internal space-charge fields and trapping positions for particles, which is enabled only by DHM, enables the determination of the charge state of these particles. By this technique, the differentiation between dielectrophoresis of neutral matter and electrophoresis of charged particles is possible. Based on this knowledge, the accumulation of charges on the surface of a z-cut lithium niobate crystal enables the electrophoretic trapping of charged particles. Due to the fact that this geometry is inherently circular symmetric, truly arbitrary patterns are possible. Furthermore, the sign of trapped particles depends on the orientation of the crystal, which facilitates the use of z-cut lithium niobate as charge sensor.

The strong internal field gradients on the surface of a $LiNbO_3$ crystal can not only trap solids suspended in liquids, but also have an influence on combinations of liquids suspended in gases, liquids in liquids or gases in liquids. In this context, dielectrophoretic forces on $LiNbO_3$ have been used to fabricate regular structures from PDMS, an initially liquid, heat-curable, transparent polymer. After the curing process, the solid modulated PDMS structures remain even if the internal space-charge fields are erased. By changing the illumination pattern and hence the electric field gradients, the polymer shape can be optically tuned towards asymmetric ratchet patterns. As an indispensable preliminary step, a method was described to measure the thickness of spin-coated polymer layer using the secondary modulation of the Fabry-Perot resonances. As a last application, lithium niobate crystals were employed as the base material to which a PDMS droplet generator was attached. Depending on the electrical properties of gas bubbles and liquids in this device, both attractive and repulsive forces could be observed. This result allows for the optically-induced sorting of droplets out of a continuous stream as well as the inscription of *virtual rails* for collinear flows, which opens up the possibility to construct novel optofluidic devices, such as an optical fluid router.

7.2 Outlook

The remarkable thing about science is that for every answer given, two new questions arise, so a lot of new ideas have been formed in the course of this thesis. It has been calculated in previous publications and been inferred from previous experiments that dielectrophoretic forces are mainly caused by electric field gradients in the immediate vicinity of the surface. However, in an upright setup, where light illuminates the photorefractive sample from below and particles are manipulated on top, only a fraction of the light actually reaches the relevant upper surface, for example 10 % of the incident light for a 600 μm thick crystal with an absorption coefficient of $\alpha = 40\,\text{cm}^{-1}$. A substantial increase in speed could thus be made simply by using a very thin crystal of 10μm, in which 96 % of the light would reach the surface. Although the handling of such samples is virtually impossible, a good compromise can be achieved by using thick pure $LiNbO_3$ crystals with a thin iron-doped photorefractive layer near the surface [1]. Such a sample would be mechanically robust, yet does only negligibly absorb light, except in the very thin surface layer. In combination with the results from the reduction technique presented in this thesis, it should be possible to push the induction time for internal field gradients to the sub-second regime, so that lithium niobate crystals become a serious alternative to glass substrates in microfluidic applications, where single droplets could then be sorted in near real-time based on their electrical properties.

The introductory part of this thesis explained that a major beneficial aspect of optical techniques is the inherent flexibility of trapping patterns. However, for the presented experiments, such a reconfigurability has not yet been achieved. This is mostly due to the fact that the charge redistribution inside the photorefractive materials is a non-instantaneous effect, which means that upon changing the incident light field, the old electric field gradients do not instantly vanish. However, with substantially faster crystals, a writing scheme can be conceived in which the gradients are erased in between two pattern-writing cycles. It has been mentioned before in Sect. 5.3 that all digital holographic measurements of the internal fields are necessarily an integration over the phase retardations induced over the whole crystal thickness. Even if special care was taken to obtain an incident light pattern that is invariant over the crystal thickness, the interaction between light field and refractive index modulations contains the danger of deteriorating the sharpness near the top surface. In a diffusion-doped crystal, space charge fields originate only from the doped regions, which improves the measurement accuracy of DHM and the fidelity of induced charged patterns.

Last but not least, the ability of $LiNbO_3$ to fabricate PDMS gratings and to manipulate liquid or gas droplets was the subject of the two distinct optofluidic applications.

Besides their optical properties, these PDMS gratings can have further mechanical applications as a stamp for ink or DNA, in which they can be dripped and subsequently be transferred to any desired substrate. The fabrication of a pre-designed micro-structured large-area surface would be accomplished in less than an hour's time due to the fact that the crystals need only to be illuminated with the desired pattern to yield the desired relief structure. In the microfluidic environment, the previously demonstrated ability to route collinear flows not only allows the deflection of air bubbles or droplets. In a collinear flow of a core liquid in a cladding liquid of lower refractive index, hence a liquid-liquid (L^2) waveguide, DEP forces allow the optical manipulation of the waveguide. Such a guiding structure can possibly have more complex forms than the patterns that can be realized by collinear flows alone [2].

References

1. M.V. Ciampolillo, A. Zaltron, M. Bazzan, N. Argiolas et al., Iron doping of lithium niobate by thermal diffusion from thin film: study of the treatment effect. Appl. Phys. A Mater. Sci. Process. **104**(1), 453–460 (2011)
2. D. Wolfe, R. Conroy, P. Garstecki, B. Mayers et al., Dynamic control of liquid-core/liquid-cladding optical waveguides. Proc. Nat. Acad. Sci. U.S.A. **101**(34), 12434–12438 (2004)

Appendix A
Phase Unwrapping

It has been shown in detail in Sect. 3.6.3.2 that the calculation of the imaginary part of an interferogram occurs as a byproduct of the demultiplexing of carrier frequency and image information and can be reduced to two Fourier transform operations. As a rule of thumb, a calculation time of less than approximately 800 ms/1.3 megapixel image on current PC hardware was given. However, this time applies only in the case of perfectly error-free phase images, where the unwrapping process takes negligible effort. This appendix will describe the general problem of two-dimensional phase unwrapping as compared to the one-dimensional case. The problem of phase ambiguity mathematically arises from the fact that the one-dimensional `arctan()` function, that calculates the phase for each pixel (cf. Eq. 3.27), yields only values ranging from $-\pi/2 \leq \varphi < \pi/2$, the two-dimensional `arctan2()`, which further considers the quadrant positions of imaginary and real part, is only defined to yield values $-\pi \leq \varphi < \pi$. Since this case is typically encountered in optical phase measurement problems, phase values will in the following be assumed to be results of the `arctan2()` function and to cover a range of 2π. Physically, this ambiguity is a result of the fact that a light wave is 2π periodic, hence the interference of light waves with a phase shift $\Delta\varphi$ larger than this value cannot be distinguished from the respective shift $\Delta\varphi' = \Delta\varphi \bmod 2\pi$. A direct consequence is that any height profile or internal phase retardation must be unwrapped to obtain continuous values.

In the one-dimensional case, Itoh's algorithm describes the solution of this problem by integration of the phase gradients from pixel to pixel [1]. In his derivation, the true unwrapped phase is calculated from

$$\varphi(m) = \varphi_0 + \sum_{n=1}^{m} \Delta(n), \tag{A.1}$$

where φ_0 is the inital phase of a reference point and $\Delta(n)$ are the wrapped phase differences, computed by differentiation of the result from the phase retrieval process and wrapping this difference back to the original π interval. This solution implies

M. Esseling, *Photorefractive Optoelectronic Tweezers and Their Applications*,
Springer Theses, DOI 10.1007/978-3-319-09318-5

that the original signal is sampled at a sample rate sufficiently high, so that the "true" phase difference—as compared to phase jumps introduced by the wrapping process—between neighboring points is always in the range of $-\pi/2 \leq \varphi(i) - \varphi(i-1) < \pi/2$. Only if this condition is true, the original phase can be accurately recovered. Extending the discrete approach of Itoh to continuous space, his algorithm equals the line integration of the phase differences [2]:

$$\varphi(\vec{r}) = \varphi_0 + \int_C \nabla\varphi d\vec{r}. \tag{A.2}$$

While in the one-dimensional case, the path of the line integral is fixed, two dimension offer the possibility to get from one point to another by a multitude of possible curves. Ghiglia and Pritt have thus reduced the problem of two-dimensional phase unwrapping to the problem of path invariance of the unwrapping [2]: Only when the line integral from one point to another is independent of the path C, a correct phase value can be calculated without significant unwrapping errors. It is known from complex analysis and can be easily understood that this is equivalent to the condition that the integral along any closed path is zero:

$$\oint f(\vec{r})d\vec{r} = \int_{b,C_1}^{a} f(\vec{r})d\vec{r} + \int_{a,C_2}^{b} f(\vec{r})d\vec{r} = \int_{b,C_1}^{a} f(\vec{r})d\vec{r} - \int_{b,C_2}^{a} f(\vec{r})d\vec{r} \stackrel{!}{=} 0 \tag{A.3}$$

It should be noted that the above criteria are also equivalent to some other conditions, but the one of vanishing closed path integrals is most easily implemented in digital image processing algorithms, as will be shown later. In the presence of phase noise, such a path integral can yield a non-zero value, also called a *phase residue charge* or *residue*, in analogy to residues encountered in complex analysis [3]:

$$\oint_C \nabla\varphi(\vec{r})d\vec{r} = \sum \text{residues} = 2\pi \times charge \tag{A.4}$$

The function $\nabla\varphi(\vec{r})$, which denotes the phase gradients in all directions, is easily computed because the phase data of DHM already exist in matrices. The simplest closed path that can be tested is along a 2×2 pixel path. So a residue of charge $\pm m$ can be attributed to the space in between four adjacent pixels, whenever the sum of all four gradients differs from zero. Note that these gradients correspond to the gradients from Eq. A.1, so all gradients in 2D must be wrapped to the interval $-\pi \leq \Delta\varphi(i, k) < \pi$. If such a residue is detected in the path, a charge of $\pm m$, depending on the magnitude of the residue is assigned to the middle of this pixel cluster. To ensure path invariance for all integration paths, the unwrapping path may never circle such a residue. To eliminate their influence on the unwrapping process, residues of opposite charge are connected (balanced) by so-called *branch cuts*, i.e.

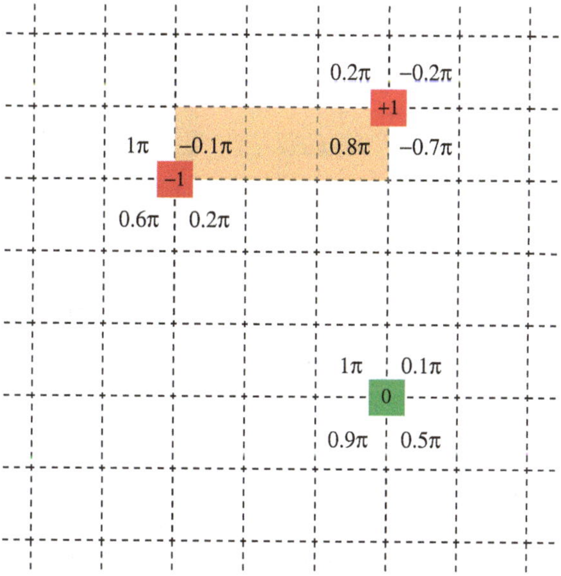

Fig. A.1 Placement of branch cuts: If the sum of the phase gradients along a 2×2 path yields a non-zero value of $2\pi \pm m$, a residue of *charge* or *polarity* $\pm m$ (*red*), otherwise a zero (*green*) is assigned to the *middle* of this pixel cluster. In order to ensure that the integration of phase gradients remains path invariant, Goldstein et al. suggested the placement of *branch cuts* (*orange*) to balance the residues

a line of pixels which must not be crossed during the unwrapping procedure. In this way, for any closed integration path, the sum of included residues will always equal zero. Residues that cannot be balanced by residue of opposite charge can be connected to the outer borders of an image matrix by a branch cut. An example for such a branch cut placement can be found in Fig. A.1. The density of residues is a direct measure for the quality of a phase image [2]. In the presence of a high number of residues, branch cuts may separate whole regions of phase information from the rest of the image. Naturally, for those regions, the correctly unwrapped phase cannot be obtained (Fig. A.2).

The final task is now to unwrap the phase data for the whole image along a path that never crosses any branch cuts. This is where the optimization of the Goldstein algorithm comes into play. A very good implementation for MATLAB exists, where no special path for unwrapping is designed, but the pixels are unwrapped one at a time [4], starting from a reference pixel and continuing the unwrapping in a closed path, also termed as *flood filling*. While this already leads to nicely unwrapped images, it is a time-consuming process since pixelwise operations are slow in MATLAB. On the other hand, the internal `unwrap()` function performs Itoh's 1D unwrapping procedure on and is highly optimized for columns or lines, but does not implement the concept of residues, so that unwrapping errors occur in two dimensions even for high-quality images with very few residues. The idea is now to combine the optimized unwrapping with Goldstein's branch cut algorithm in the following procedure:

1. Calculate the phase gradient matrices $\nabla_x \varphi = \varphi(i, k) - \varphi(i-1, k)$ and $\nabla_y \varphi = \varphi(i, k) - \varphi(i, k-1)$.
2. Calculate the sum of those gradients along a closed loop for all 2×2 pixel clusters.

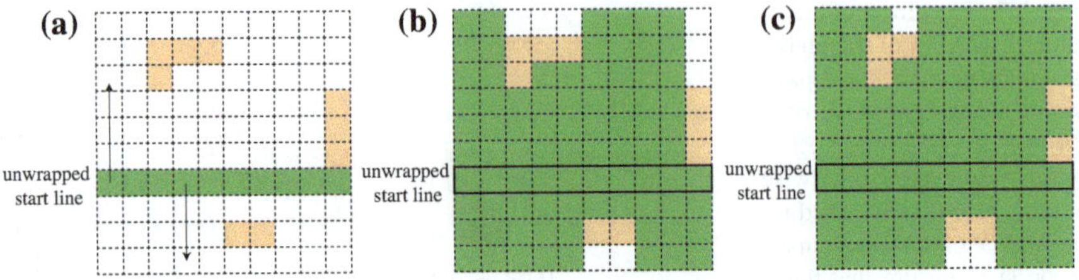

Fig. A.2 2D phase unwrapping with the optimized Goldstein algorithm: an *inital line*, which avoids branch cuts (*orange*) is unwrapped; from these unwrapped values, the unwrapping of columns can be continued in both directions using MATLAB's optimized 1D unwrapping algorithm; the rest of the pixels as well as the branch cuts are unwrapped using pixelwise unwrapping, called *flood filling*. **a** line unwrapping, **b** unwrapping of columns, **c** flood fill

3. If for pixel cluster the sum is non-zero, assign a residue of charge $\pm m$ to the space in between those pixels.
4. Place branch cuts to balance residues or to connect to the borders of the image. These first 4 steps can make use of the already very good code from [4].
5. Look for a line/column with no branch cuts at all, or as little branch cuts as possible.
6. Use the internal `unwrap()` function to quickly unwrap this line/column completely or until the branch cut using Itoh's 1D algorithm.
7. Starting from this unwrapped line/column, use MATLAB's `unwrap()` function to unwrap all perpendicular columns/lines to both sides completely or until a branch cut has been reached.
8. Determine the phase value of all pixels that have not been unwrapped by simple flood filling.

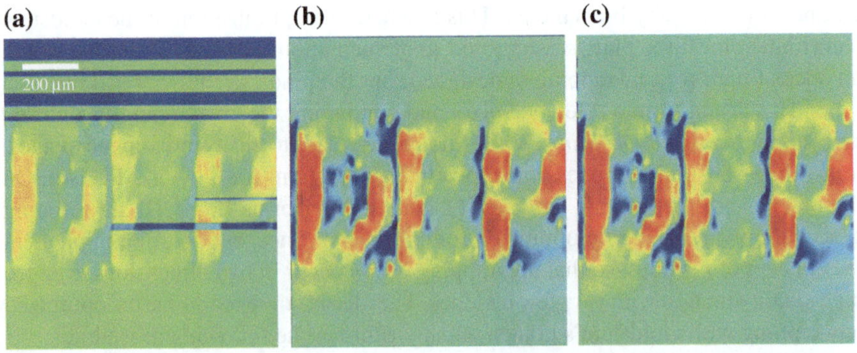

Fig. A.3 Benchmark of three unwrapping algorithms: the 1D Itoh algorithm is by far the fastest, but unreliable; the Goldstein algorithm is robust and precise, but slow; the optimized Goldstein algorithm for MATLAB combines both advantages and yields a quality comparable to the Goldstein at significantly reduced time. **a** Itoh: $t = 0.18$ s. **b** Goldstein: $t = 158$ s. **c** optimized Goldstein: $t = 0.87$ s

Even with this simple approach, a significant increase in speed can be achieved as depicted in Fig. A.3. It shows the background corrected unwrapping results of a hologram that was acquired after the letters DEP have been induced in a photorefractive $LiNbO_3$ crystal. The performance of the three algorithms described above can be assessed in terms of image quality and unwrapping speed. In terms of speed, the extension of Itoh's one-dimensional algorithm to two dimensions is by far the fastest, but serious unwrapping errors can be expected for every residue that is crossed during the unwrapping process. The simplest Goldstein flood-fill implementation is very robust and results in a perfectly unwrapped phase image except for the residues and branch cuts, where minor phase unwrapping errors cannot be avoided. The same qualitative result is obtained by the optimized Goldstein algorithm in less than 1 % of the calculation time, simply by making use of the optimized internal unwrapping algorithm.

As a concluding remark, it should be noted that the optimized algorithm presented is faster only for programming languages that possess an optimized version of the Itoh algorithm. Its performance also is greatly dependent on the structure and density of the residues. In a worst-case scenario, where there are only very lines/columns without branch cuts, its performance approaches that of the conventional flood filling Goldstein procedure. Apart from the previously described methods, two-dimensional phase unwrapping is still a vital field of research and numerous methods have been developed, such as quality-guided unwrapping, where the quality of the phase information of each pixels is rated with respect to the neighboring pixels by their phase gradient fluctuation or their correlation function. Subsequently, only pixels with a quality above a certain threshold will be unwrapped [2, 5]. In contrast to these algorithms, which all belong to the *path-following class*, other methods employ a more general formulation where each individual pixel is varied so that the a predefined error norm is minimized. For a thorough description, the reader is referred to a very comprehensive book about two-dimensional phase unwrapping by Ghiglia and Pritt [2].

References

1. K. Itoh, Analysis of the phase unwrapping algorithm. Appl. Opt. **21**(14), 2470 (1982)
2. D.C. Ghiglia, M.D. Pritt, *Two-Dimensional Phase Unwrapping: Theory, Algorithms and Software* (Wiley, New York, 1998)
3. R. Goldstein, H. Zebker, C. Werner, Satellite radar interferometry—two-dimensional phase unwrapping. Radio Sci. **23**(4), 713–720 (1988)

4. B. Spottiswoode, Goldsteinunwrap2d for MATLAB (2008), http://www.mathworks.com/matlabcentral/fileexchange/22504-2d-phase-unwrapping-algorithms/content/GoldsteinUnwrap2D.m
5. D. Bone, Fourier fringe analysis—the 2-dimensional phase unwrapping problem. Appl. Opt. **30**(25), 3627–3632 (1991)

Appendix B
Building Microstructures from Polydimethylsiloxane

In many of the optofluidic applications for photorefractive optoelectronic tweezers, the polymer *polydimethysiloxane* (PDMS) plays a key role, either as a means to construct microchannels or as the building material to directly fabricate diffraction gratings on top of a $LiNbO_3$ substrate. It is therefore convenient to summarize the basic properties that have made (and still keep) PDMS the material of choice for a variety of microfluidic applications [1–3] (Fig. B.1).

PDMS is a viscoelastic polymer that is liquid in its non-crosslinked state. It is shipped and typically used with a cross-linking chemical developer. When mixed at the recommended composition, the mixture gradually solidifies as the siloxane chains become more and more cross-linked. For Dow Corning Sylgard 184®, the PDMS kit used most often and also in this thesis, the mixing recommendation is a 10:1 ratio (by weight) between prepolymer and cross-linker [4]. The cross-linking process takes up to 48 h at room temperature, but can be significantly sped up to 10 min by raising the ambient temperature to 150 °C. It was found during the experiments in this thesis that curing the mixed polymer at reduced temperatures of up to 100 °C results in a slightly better optical quality due to reduced striations in the material, possibly induced by mechanical stress in the material if it is cured too fast. The beneficial properties of PDMS that qualify it for microfluidic and optofluidic applications are its very good optical transmittance over the whole visible spectrum and the fact that it is non-toxic, chemically inert and electrically isolating [5, 6]. Due to its initially liquid state, it can be easily used for soft lithography where it is shaped into a variety of desired forms. On top of all these aspects, PDMS building blocks are reversibly self-adherent by van-der-Waals forces, so for most experiments at moderate pressures, no additional sealing is necessary [5]. This is especially important for the applications presented in this thesis since photorefractive crystals are not a disposable. Despite the fact that industrial scale production has led to a significant decrease in the costs of $LiNbO_3$ wafers, a permanent connection of PDMS channels to a photorefractive crystal is not desired. In other cases where a sealing of PDMS to the supporting glass or PDMS structure is necessary, permanent bonds have been demonstrated by exposing the surfaces to oxygen plasma or strong UV illumination prior to connecting them [7]. In its cured, unmodified state PDMS is hydrophobic

M. Esseling, *Photorefractive Optoelectronic Tweezers and Their Applications*,
Springer Theses, DOI 10.1007/978-3-319-09318-5

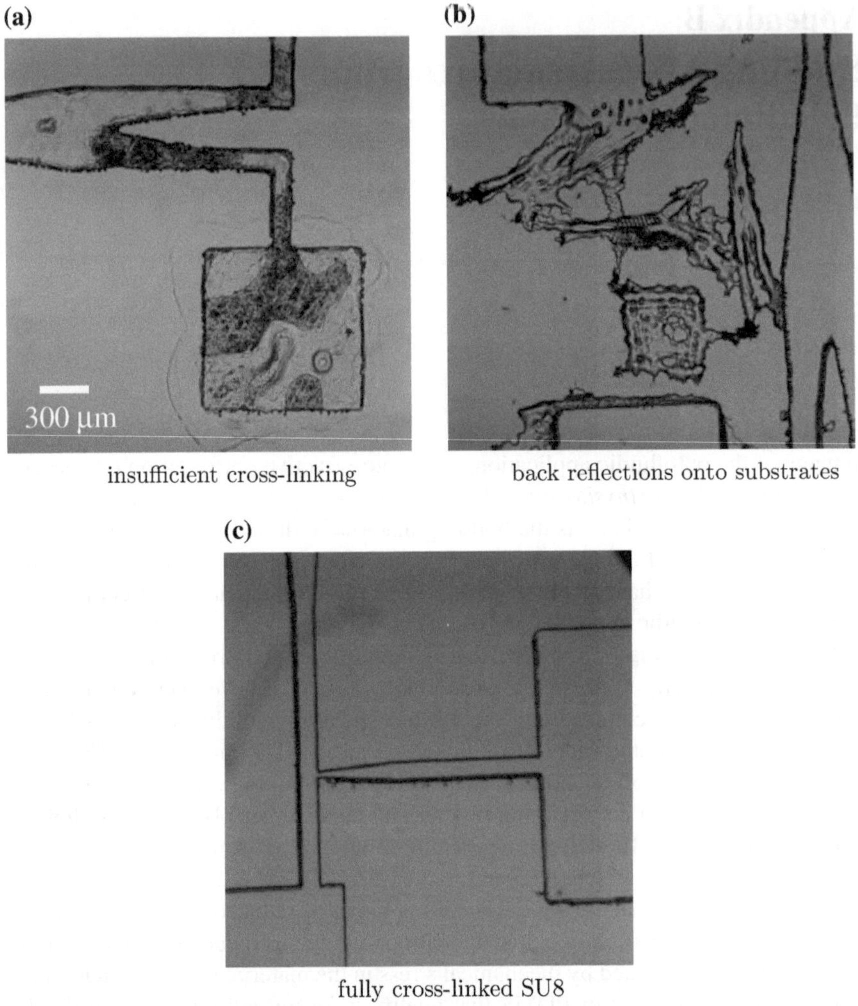

Fig. B.1 Typical mistakes during SU8 master fabrication: **a** insufficient cross-linking, either caused by insufficient illumination dose or too short post-bake time; **b** back-reflections of the illumination pattern from lenses or camera chips can lead to the formation of demagnified replicas in the SU8; the avoidance of these errors yields a fully cross-linked SU8 master mold structure (**c**)

with a contact angle of approximately 110°, which can be reduced by a UV/ozone treatment to be less than 10° [8]. The hydrophilicity facilitates the filling of microfluidic channels but also increases the solvent resistivity against unpolar solvents, such as many of the substances used in this thesis. A typical challenge when dealing with PDMS microchannel and apolar solvents is that the solvents diffuse into the bulk PDMS and induce swelling and severe deformations, which render the fabricated

Table B.1 SU8 film thickness d after postbake for SU8-100 diluted with 10 wt.% acetone and spin-coated for $t = 60$ s at different spinning velocities

f/rps	d/μm
10	122
20	65
40	35
60	19

structures useless [9]. Berdichevsky et al. could show that the adsorption of apolar solvents is significantly reduced after oxidation of the surface with UV/ozone treatment [10].

Master Fabrication for PDMS Replica Molding

In order to form the shape of PDMS according to the needs of a special applications, it is typically cured in an appropriately shaped mold. In the easiest form, for example to produce a simple reservoir, such a mold can be constructed from adhesive tape as the barrier for liquid PDMS and a metal cylinder sitting in the middle. After curing, the metal cylinders is removed and the peeled-off PDMS forms a self-adherent reservoir that can be applied to any smooth surface, such as a $LiNbO_3$ crystal. However, for most applications, a more sophisticated design, such as a micro channel, micro mixer or droplet generator is required. Such a master can be prepared by preparing a lithographic mask, transferring it to a silicon wafer and etching all the non-covered parts, so that the predesigned relief structure remains. Depending on the etching time, the height of the structures can be carefully controlled. A silicon wafer is arguably the most robust solution of constructing a master mold, since the microchannel relief and the bottom substrate are produced from a single crystalline block of material and can be used as casting mold over and over again with almost no degradation. Conversely, for many applications in optofluidics laboratories, where experiments are typically carried out as proof-of-concept experiments with little repetition, a high flexibility of the fabricated master is more important than a high robustness. In this case, the fabrication of an SU8 master structure is favorable, since it can be accomplished with typical lab equipment, requiring only substances of limited toxicity. SU8 is a UV-curable photoresist that has found wide applications in the fabrication of micro-electromechanical devices (MEMS). This part of the appendix does not aim to describe the protocol of fabricating SU8 master structures, because such a protocol as well as the typical properties of SU8 have been described in great detail elsewhere (see [11] and references therein). This section should rather give some additional technical advice and lab expertise, which has proven useful in the preparation of the droplet generator described in Sect. 6.4.

The positive photoresist SU8 is a photopolymer, in which ultraviolet light can create acids that, in a subsequent heat treatment, polymerize the material and lead to the formation of solid structures that cannot be washed away by solvents such as the proprietary developer *mr-Dev 600* (2-Methoxy-1-methylethylacetat) [12] or acetone. Before the actual illumination can take place, the SU8 must be appropriately coated to a glass substrate and pre-treated using the following process steps:

1. Thoroughly clean the glass surface mechanically and chemically using acetone. Then pre-treat the surface with proprietary HDMS 80/20 primer before applying the SU8. This step is crucial for a good adhesion of the cured SU8 to the substrate. Omitting this first step will most likely result in master structures that adhere better to the PDMS mold than to the glass substrate, hence are peeled off from the glass in the first casting process. To pre-treat the surface, first bake the cleaned substrate for 3 min to remove all remaining liquid. Then apply a drop of HDMS and let it etch the glass for a minimum time of 1 min. After that, spin-dry the substrate in a spin-coater.
2. Apply a drop of SU8 resist and spin-coat to the desired thickness. Different SU versions are produced with different viscosities leading to a variety of film thicknesses [12]. In case the available version of SU8 (SU8-100 in this case) is too viscous, it can be diluted with acetone for the easier application of more homogeneous and thinner films. Table B.1 shows the thicknesses for SU8 diluted with 10 wt.% acetone and spin-coated for 60 s. It was found that the SU8 can be diluted by up to 15 wt.% of acetone.
3. After the spin-coating process, the SU8 films must be pre-baked to evaporate the remaining solvent. Insufficient pre-bake will result in the movement of SU8 during illumination, hence distorted structures. A pre-bake of 10 min at 65 °C, followed by 30 min at 95 °C has been found to yield good fidelity of the SU8 and sufficient adhesion to the substrate. Longer pre-baking times are also possible, since the curing starts only after the incidence of UV light. In order to prohibit premature curing, the unexposed samples should be kept away in the dark if they are stored for a longer time.
4. After the pre-bake, the samples are ready for illumination. Several dosage guidelines for different thicknesses and wavelengths exist, but at the wavelength of 405 nm, which was used in this thesis, the optimal illumination time for a 65 μm thick film was found to be 4 min at an mean intensity of $110\,\mathrm{mW\,cm^{-2}}$. During illumination, special care should be taken that the light is totally absorbed after the sample. Back-reflections from camera chips, lenses or glass slides can induce additional detrimental points of photo-initiation in the SU8.
5. For the final curing, the sample should be post-baked at 65 °C for 5 min, then 25 min at 95 °C. As before, the slow ramping of the temperature avoids the induction of stress in the sample. Therefore, the sample should also be cooled down moderately, preferably on the hotplate, instead of adding solvent to cool it down. Cooling or heating the sample too fast most likely will result in cracks in the SU8 structure.
6. When the sample is cooled down, the remaining non-cured polymer is washed away adding the proprietary developer *mr-Dev 600* from *microResist Technology*, Berlin. The sample should be kept in the developer for several minutes, allowing

for all remains to be washed away. To remove the rest of the developer as well, a final rinse with acetone was found to be efficient.

7. Check the quality of the SU8 master under a microscope. In the case of remaining polymer, repeat the previous step. When the quality is sufficient, a final optional post-post-bake (sic!) at 150 °C for 30 min can be performed to increase the adhesion to the glass. Using this final bake, molds have been obtained that could be reused for more than 25 casting processes. As always, slow heating and cooling is a prerequisite for crack-free SU8 structures.

Following these simple rules, it is possible to make a very robust SU8 casting mold within 3 h. This has the advantage that multiple parameters can be tested in parallel, since the illumination mask can be exchanged for each new mold. The illumination masks were produced by femtosecond laser ablation of an aluminum-coated glass substrate, which has the advantage of high absorbance in the opaque regions and very high resolution. A similar pattern can be achieved by using an amplitude spatial light modulator that is demagnified onto the SU8 sample. However, special attention should be paid to the contrast between dark and bright regions that is achieved by the SLM.

References

1. E. Leclerc, Y. Sakai, T. Fujii, Microfluidic PDMS (polydimethylsiloxane) bioreactor for large-scale culture of hepatocytes. Biotechnol. Prog. **20**(3), 750–755 (2004)
2. M. Zhang, J. Wu, L. Wang, K. Xiao et al., A simple method for fabricating multi-layer PDMS structures for 3D microfluidic chips. Lab Chip **10**(9), 1199–1203 (2010)
3. W. Song, D. Psaltis, Pneumatically tunable optofluidic 2 × 2 switch for reconfigurable optical circuit. Lab Chip **11**(14), 2397–2402 (2011)
4. Dow Corning, Product information: Sylgard 184 silicone elastomer (2013)
5. J. McDonald, G. Whitesides, Poly(dimethylsiloxane) as a material for fabricating microfluidic devices. Acc. Chem. Res. **35**(7), 491–499 (2002)
6. D.K. Cai, A. Neyer, R. Kuckuk, H.M. Heise, Optical absorption in transparent PDMS materials applied for multimode waveguides fabrication. Opt. Mater. **30**(7), 1157–1161 (2008)
7. S. Bhattacharya, A. Datta, J. Berg, S. Gangopadhyay, Studies on surface wettability of poly(dimethyl) siloxane (PDMS) and glass under oxygen-plasma treatment and correlation with bond strength. J. Microelectromech. Syst. **14**(3), 590–597 (2005)
8. K. Efimenko, W. Wallace, J. Genzer, Surface modification of Sylgard-184 poly(dimethyl siloxane) networks by ultraviolet and ultraviolet/ozone treatment. J. Colloid Interface Sci. **254**(2), 306–315 (2002)
9. J. Lee, C. Park, G. Whitesides, Solvent compatibility of poly(dimethylsiloxane)-based microfluidic devices. Anal. Chem. **75**, 6544–6554 (2003)

10. Y. Berdichevsky, J. Khandurina, A. Guttman, Y. Lo, UV/ozone modification of poly(dimethylsiloxane) microfluidic channels. Sens. Actuators B Chem. **97**(2–3), 402–408 (2004)
11. Memscyclopedia: Su-8: Thick photo-resist for mems (2013), http://memscyclopedia.org/su8.html
12. microResist technology GmbH Homepage (2013), http://www.microresist.de/

Curriculum Vitae

Personal Data

Name:	Michael Esseling
Date of birth:	01.10.1984
Place of birth:	Stadtlohn
Nationality:	German
Parents:	Josef Esseling, Marianne Esseling (born Gajewiak)

School Education

1991–1995	Hordtschule, Stadtlohn (Primary School)
1995–2004	Geschwister-Scholl-Gymnasium, Stadtlohn (High School)
May 2004	Allgemeine Hochschulreife (University entry exam)

University Education

2004–2010	Studies of Physics at the Westfälische Wilhelms-Universität Münster
02.03.2010	Diploma degree; diploma thesis "Applications for Photorefractive Crystals in Microfluidics"; awarded the *Infineon Master Award* as best diploma thesis of the semester
03/2010–04/2014	Ph.D. studies in the *Nonlinear Photonics* workgroup; supervisor Prof. Dr. Cornelia Denz

Peer-Reviewed Publications

1. *Two-dimensional dielectrophoretic particle trapping in a hybrid PDMS/crystal system*, M. Esseling, F. Holtmann, M. Woerdemann, and C. Denz, *Optics Express* **18**, 17404–17411 (2012)
2. *Depth-resolved velocimetry of Hagen-Poisseuille and Electroosmotic Flow using Dynamic Phase-Contrast Microscopy*, M. Esseling, F. Holtmann, M. Woerdemann, and C. Denz, *Applied Optics* **49**, 6030–6038 (2010)

M. Esseling, *Photorefractive Optoelectronic Tweezers and Their Applications*,
Springer Theses, DOI 10.1007/978-3-319-09318-5

3. *Multimodal biophotonic workstation for live cell analysis*, M. Esseling, B. Kemper, M. Antkowiak, D.J. Stevenson, L. Chaudet, M.A.A. Neil, P.W. French, G. V. Bally, K. Dholakia, and C. Denz, *Journal of Biophotonics* **5**, 9–13 (2012)
4. *Holographic optical bottle beams*, C. Alpmann, M. Esseling, P. Rose, and C. Denz, *Applied Physics Letters* **100**, 111101 (2012)
5. *Opto-electric particle manipulation on a bismuth silicon oxide crystal*, M. Esseling, S. Glaesener, F. Volonteri, and C. Denz, *Applied Physics Letters* **100**, 161903 (2012)
6. *Multiplexing and switching of virtual electrodes in optoelectronic tweezers based on lithium niobate*, S. Glaesener, M. Esseling, and C. Denz, *Optics Letters* **37**, 3744–3746 (2012)
7. *Photophoretic trampoline - Interaction of single airborne absorbing droplets with light*, M. Esseling, P. Rose, C. Alpmann, and C. Denz, *Applied Physics Letters* **101**, 131115 (2012)
8. *Dynamic Light Cages: Putting Absorbing Matter Behind Bars*, C. Alpmann, M. Esseling, P. Rose, and C. Denz, *Optics and Photonics News: Optics In 2012* **23**, 48 (2012)
9. *Advanced optical trapping by complex beam shaping*, M. Woerdemann, C. Alpmann, M. Esseling, and C. Denz, *Laser & Photonics Reviews* **7**, 839–854 (2013)
10. *Highly reduced iron-doped lithium niobate for optoelectronic tweezers*, M. Esseling, A. Zaltron, N. Argiolas, G. Nava, J. Imbrock, I. Cristiani, C. Sada, and C. Denz, *Applied Physics B* **113**, 191–197 (2013)
11. *Charge sensor and particle trap based on z-cut lithium niobate*, M. Esseling, A. Zaltron, C. Sada, and C. Denz, *Applied Physics Letters* **103**, 061115 (2013)

Selected Conference Presentations

Oral Presentations

1. *Two-dimensional dielectrophoretic particle trapping in a hybrid PDMS/crystal system*, M. Esseling, F. Holtmann, M. Woerdemann, and C. Denz, *EOS Annual Meeting*, **Paris (France)**, 26–28.10.2010
2. *Microfluidic particle manipulation on electro-optic surfaces*, M. Esseling, S. Glaesener, and C. Denz, *OSA optical Trapping Applications*, **Monterey (USA)**, 04–06.04.2011
3. *Particle Manipulation on electrooptic surfaces*, M. Esseling, S. Glaesener, and C. Denz, *Nonlinear Materials and Applications (NOMA)*, **Cetraro (Italy)**, 06–0.06.2011
4. *Opto-electric particle manipulation on a bismuth silicon oxide surface*, M. Esseling, H. Futterlieb, S. Glaesener, F. Volonteri, and C. Denz, *SPIE Optical Trapping & Optical Micromanipulation IX*, **San Diego (USA)**, 12–16.08.2012

5. *The photophoretic trampoline*, M. Esseling, P. Rose, C. Alpmann, and C. Denz, *EOS Optofluidics*, **Munich (Germany)**, 13–15.05.2013; awarded **EOS Best Oral Presentation Award**
6. *Opto-electronic tweezers based on reduced lithium niobate*, M. Esseling, A. Zaltron, N. Argiolas, G. Nava, J. Imbrock, I.Cristiani, C.Sada, and C. Denz, *Photorefractives 14*, **Winchester (UK)**, 04–06.09.2013

Poster Presentations

1. *Two-dimensional dielectrophoretic particle manipulation*, M. Esseling, F. Holtmann, M. Woerdemann, and C. Denz, *COST Summer School*, **Visegrad (Hungary)**, 05–08.10.2010
2. *Photorefractive Optoelectric Tweezers*, M. Esseling, S. Glaesener, and C. Denz, *EOS Optofluidics*, **Munich (Germany)**, 23–25.05.2011

